THE ONE

ALSO BY HEINRICH PÄS

The Perfect Wave

THE ONE

HOW AN ANCIENT IDEA HOLDS THE FUTURE OF PHYSICS

HEINRICH PÄS

BASIC BOOKS
New York

Copyright © 2023 by Heinrich Päs
Cover design by Chin-Yee Lai
Cover images © KAWEESTUDIO / Shutterstock.com; © redstone / Shutterstock.com; © Bankrx / Shutterstock.com
Cover copyright © 2023 by Hachette Book Group, Inc.
Illustrations by Frigga Päs

Hachette Book Group supports the right to free expression and the value of copyright. The purpose of copyright is to encourage writers and artists to produce the creative works that enrich our culture.

The scanning, uploading, and distribution of this book without permission is a theft of the author's intellectual property. If you would like permission to use material from the book (other than for review purposes), please contact permissions@hbgusa.com. Thank you for your support of the author's rights.

Basic Books
Hachette Book Group
1290 Avenue of the Americas, New York, NY 10104
www.basicbooks.com

Printed in the United States of America

First Edition: January 2023

Published by Basic Books, an imprint of Perseus Books, LLC, a subsidiary of Hachette Book Group, Inc. The Basic Books name and logo is a trademark of the Hachette Book Group.

The Hachette Speakers Bureau provides a wide range of authors for speaking events. To find out more, go to www.hachettespeakersbureau.com or email HachetteSpeakers@hbgusa.com.

The publisher is not responsible for websites (or their content) that are not owned by the publisher.

Wheeler's U was first published in Zurek 1990, p. ix. Reprinted with permission of the estate of John Archibald Wheeler.

Print book interior design by Jeff Williams.

Library of Congress Cataloging-in-Publication Data
Names: Päs, H. (Heinrich), author.
Title: The one : how an ancient idea holds the future of physics / Heinrich Päs.
Description: First edition. | New York, NY : Basic Books, 2023. | Includes bibliographical references and index.
Identifiers: LCCN 2022019806 | ISBN 9781541674851 (hardcover) | ISBN 9781541674844 (ebook)
Subjects: LCSH: Physics—Philosophy. | Monism. | One (The One in philosophy) | Quantum theory. | Cosmology.
Classification: LCC QC6 .P2965 2023 | DDC 523.101—dc23/eng20220927
LC record available at https://lccn.loc.gov/2022019806

ISBNs: 9781541674851 (hardcover), 9781541674844 (ebook)

LSC-C

Printing 2, 2023

For Sara
You are the One for me

CONTENTS

Introduction: Stargazing — 1
1. The Hidden One — 9
2. How All Is One — 39
3. How One Is All — 65
4. The Struggle for One — 103
5. From One to Science and Beauty — 147
6. One to the Rescue — 191
7. One Beyond Space and Time — 219
8. The Conscious One — 249
Conclusion: The Unknown One — 273

Acknowledgments — 291
Further Reading — 293
Glossary — 299
Notes — 305
Bibliography — 327
Index — 343

From all things One and from One all things.
—HERACLITUS

There are many indications that, following the recursive pattern of scientific revolutions, we are now witnessing the beginning of the phase of crisis . . . This is the most complex and intense moment of scientific research, when revolutionary and unprejudiced ideas are needed for a real paradigm change.
—GIAN GIUDICE, HEAD OF THE
THEORETICAL PHYSICS DEPARTMENT, CERN

To lend wings to physics once again.
—FRIEDRICH WILHELM JOSEPH SCHELLING

INTRODUCTION

Stargazing

ONE VERY EARLY MORNING IN MID-OCTOBER 2009, I was waiting alone in a deserted and pitch-dark alley in San Pedro, in the middle of the Chilean Atacama Desert, one of the driest spots on earth. Above me, countless stars were sparkling, so mesmerizing that I struggled to keep an eye out for the tour guide's truck, coming to pick me up for a trip up to the Altiplano to watch the flamingoes stalking a secluded salt flat in the first light of the rising sun. Never before or after have I seen a more magnificent sky, though there have been other, similarly magical moments: counting shooting stars from the deck of a sailboat while crossing the Baltic Sea, practicing full-moon surfing off Waikiki Beach in Hawaii, or stepping out of a ski cabin at night, halfway up a mountain in the Austrian Alps, only to be stopped in my tracks by the bright band of the Milky Way's galactic disk. In such moments, I have felt entirely small and insignificant and yet, at the same time, strangely at home in the universe.

But what does it mean to feel at home in the universe? What do we actually mean when we talk about the "universe"? Etymologically, the word comes from the Latin *universum*, meaning something like "all things combined together into one." Yet, when we speak of the universe, we usually refer to outer space, our cosmic environment, stars, planets, galaxies, a vast realm filled with countless objects. Apparently, what we refer to as the "universe" and what the term actually means have little in common, if anything at all.

Almost all the celestial objects you can identify in the night sky belong to our own galaxy, the Milky Way, which in total hosts more than one hundred billion stars. And the Milky Way itself is only one among about a trillion galaxies. As impressive as these numbers are, these visible objects are only the tiniest part of the entire universe. For every star you can spot out there, there exists about ten times more mass in nonluminous matter, such as gas clouds billowing around in interstellar space. Even more so, for all ordinary matter there exists five times as much mass in "dark matter," expected to be made of exotic, unknown particles floating across the universe. And finally, there exists three times more "dark energy," the puzzling fuel that drives the fabric of space-time to expand faster and faster.

So much for our "universe."

But according to modern cosmology, maybe even our universe is not everything—there may be more than just a single universe. Cosmologists now describe an epoch of accelerating expansion in the very early times, called "cosmic inflation." The inflationary period terminates in a hot plasma, which we can identify with the Big Bang. But nobody knows what happened before inflation. Was there an absolute beginning? Or did inflation go on forever, and is it maybe still going on outside our own universe, in other regions of a "multiverse"? In that case it may continue to produce innumerable other "baby universes," popping up in an eternally inflating space. This, actually, is quite possible.

But that is not enough to account for "everything" either. Not even close! Beyond parallel universes, dark energy, dark matter, and trillions of galaxies with a hundred billion stars each, there may yet exist a realm of infinite possibilities, where everything that, in principle, could exist actually *does*. There you would find innumerable copies of yourself, my cat, your dog, the flamingoes of the Altiplano, everyone, of all stars and galaxies and everything mentioned above. These parallel realities are the different branches of Hugh Everett's infamous "many worlds" interpretation of quantum mechanics. In fact, they constitute another—arguably more fundamental—layer of multiverse. Increasingly more physicists are willing to accept now that they are an inherent prediction of quantum mechanics—that a functional notion of quantum mechanics is increasingly difficult to sustain without "many worlds."

And even this is not the end of the story. In addition to these parallel worlds, the quantum world contains infinite arbitrary "superpositions" of these realities. These are realities in which cats are half dead and half alive and where you are not either sitting in a chair and reading a book in the United States or driving a rental car through Europe but where both activities and places are mixed up in a way implying one cannot decide which one is true. The quantum realm encompasses everything that could be and all possible blends of these would-be realities.

Yet, standing there, under the stars, I still felt that feeling that many humans have shared: that I was somehow one with the vastness beyond myself. Is there a more daring, courageous, and flat-out overwhelming thought than to conceptualize "the whole material world," everything from "celestial bodies" to "life upon the earth" and from the "nebula stars to the mosses on the granite rocks," as the great German naturalist and discoverer Alexander von Humboldt portrayed the universe, as "One"?[1]

It seems bizarre to believe all of this could be connected. It sounds like a fairy tale fabricated by mystics or madmen. Yet the conviction that the universe is all "one" and the experience that it is comprised of many things have been an enduring conflict for humanity since its earliest days. "From all things One and from One all things": twenty-five hundred years ago, the Greek philosopher Heraclitus had expressed the thought of an all-encompassing universe in its most radical way.[2] This notion that there is but one object in the universe, the universe itself, is known to philosophers as "monism," from the ancient Greek *monos*, meaning "unique." It has inspired Plato's dialogues, Botticelli's painting *The Birth of Venus*, Mozart's opera *The Magic Flute*, and a major part of Romantic poetry from Goethe to Coleridge and Wordsworth. It has traveled with James Cook's ships around the world and driven several of the founding fathers of the United States of America, even making its way into the US Declaration of Independence as "Nature's God." The One has had such an influence on the world of ideas, on the arts and humanities, that its importance as a scientific concept is often overlooked. Taken at face value though, the hypothesis that "all is One" isn't a statement about God, spirits, or subjective mental states; it is a statement about nature, about the particles, planets, and stars out there.

As a theoretical physicist, for the past twenty-five years I have worked to figure out how tiny particles compose the world. Particles have thrilled me since the very first time I heard about them. Yet, fascinating as they are, what truly captivated me about these particles is how they can serve as a tool to uncover the foundations of reality. "What is everything made of?" was a question that started to occupy me when I still was in high school. This fascination was what got me into physics, earning me a PhD and finally a professorship. Particles kept me going when I was struggling with math, incomprehensible language, and feelings of inferiority. And particles were the driving force of my work when, over the following decades, I published more than eighty papers in refereed journals, when I wrote a *Scientific American* cover feature that got reprinted next to a piece by Stephen Hawking, and when my research made it three times onto the cover of *New Scientist* magazine. Of course, I'm not alone in this endeavor. I'm just a modest contributor in a global enterprise. There are some ten thousand researchers all over the world, including some of the most brilliant minds on the planet, working restlessly to find out how particles ultimately constitute what we see around us.

Now I believe we are on the wrong track.

Don't get me wrong. Science's most important task is to predict and explain the outcome of experiments, observations, and events. And particle physics does that with an unrivaled accuracy. Starting with a set of equations that fits onto a coffee mug, particle physicists predict the results of their experiments with a precision that would correspond to knowing the distance between London and Berlin up to less than a millimeter. But while particle physics is still more precise than any other discipline in science, it doesn't tell the full story. Because if we pay attention to the full story, we will see that particles do not compose the world; it is the other way around.

Ever since the discovery of the atom, physicists adhered to the philosophy of reductionism. According to this idea, nature could be grasped in a unified understanding by decomposing everything around us into pieces made up from the same tiny constituents. According to this common narrative, everyday objects such as chairs, tables, and books are made of atoms, atoms are composed of atomic nuclei and electrons, atomic nuclei contain protons and neutrons, and protons and neutrons consist of quarks. Elementary particles such as quarks or electrons are understood as the fundamental building blocks of the universe. Over the past fifty years, to work out and concretize this view, hundreds of thousands of pages have been filled with sophisticated equations full of strange symbols. To test these ideas, gigantic particle smashers have been built, tubes many miles long and worth billions of dollars, to accelerate subatomic matter close to the speed of light, let it crash together with violent impact, and search for even smaller or as-yet undiscovered pieces. With the help of NASA and the European Space Agency, engineering marvels have been launched into space to eavesdrop on the earliest incidents in the universe to understand how the world looked when it was but a soup of hot particles.

This philosophy has been tremendously successful, but there is a blind spot. Atoms, protons and neutrons, electrons and quarks are described by quantum mechanics. And according to quantum mechanics, it is, in general, impossible to decompose an object without losing some essential information. Particle physicists strive for a fundamental description of the universe, one that discards no information. But if we take quantum mechanics seriously, this implies that, on the most fundamental level, nature cannot be composed of constituents. The most fundamental description of the universe has to start with the universe itself.

Like any other professional physicist, I work with quantum mechanics on a daily basis. We use quantum mechanics to calculate and predict the results of the experiments, observations, and problems that interest us, be it particle collisions in giant accelerators, scattering processes in the primordial plasma of the early universe, or the behavior of electric or magnetic fields in a solid-state lab experiment. But while we almost always adopt quantum mechanics to describe specific observations and experiments, we usually don't apply it to the entire universe.

This has a mind-boggling consequence. As I will argue in this book, once quantum mechanics is applied to the entire cosmos, it uncovers a three-thousand-year-old idea: that underlying everything we experience there is only one single, all-encompassing thing—that everything else we see around us is some kind of illusion.

Admittedly, the claim that "all is One" doesn't sound like an ingenious scientific concept. On a first glance, it sounds absurd. Just look out the window. Most of the time there will be more than one car in the street. It takes two persons (at least!) for a love affair, "two or three" believers are required to hold a Mass, and twenty-two players are needed for a proper soccer game. Ages ago, astronomers convinced us that Earth is not the only planet in the universe, and today modern cosmology knows virtually innumerable stars.

But quantum mechanics changes everything. In quantum systems, objects get so completely and entirely merged that it is impossible to say anything at all about the properties of their constituents anymore. This phenomenon is known as "entanglement," and while it was pointed out by Albert Einstein and collaborators some eighty years ago, it is only now getting fully appreciated. Apply entanglement to the entire universe and you end up with Heraclitus's dogma "From all things One."

"Hold on," you may object. "Quantum mechanics applies only to tiny things: atoms, elementary particles, maybe molecules. Applying it to the universe doesn't make sense." You will be surprised to learn that there are increasingly many good hints that this conviction is wrong. Between 1996 and 2016 alone, six Nobel Prizes were awarded for so-called macroscopic quantum phenomena. Quantum mechanics seems to apply universally, a finding whose consequences are just starting to be explored.

You may throw up your hands and protest that such a discussion is pointless. Physics seems to work just fine without any such metaphysical pondering. Fact is, it doesn't. At present, physics is facing a crisis that forces us to reconsider what we understand as "fundamental" in the first place. Right now, the most brilliant particle physicists and cosmologists are alienated by experimental findings of extremely unlikely coincidences that so far defy any explanation. At the same time, the quest for a theory of everything is bereaving physics of its foundational concepts, such as matter, space, and time. If these are gone, what remains?

Quantum cosmology implies that the fundamental layer of reality is made neither of particles nor of tiny, vibrating, one-dimensional objects known as "strings," but the universe itself—understood not as the sum of things making it up but rather as an all-encompassing unity. As I will argue, this notion that "all is One" has the potential to save the soul of science: the conviction that there is a unique, comprehensible, and fundamental reality. Once this argument holds sway, it will turn our quest for a theory of everything upside down—to build up on quantum cosmology rather than on particle physics or string theory (currently the most popular candidate for a quantum theory of gravitation). Such a concept further implies the need to understand how it is possible that we experience the world as many things if everything is "One," after all. This is ensured by a process known as "decoherence," which is essential to virtually any branch of modern physics. Decoherence is the agent protecting our daily-life experience from too much quantum weirdness. And it realizes the rest of Heraclitus's tenet: "from One all things."

As a consequence, we will have to work out how such a notion changes our perspective on philosophy's deepest questions—"What is matter?" "What is space?" "What is time?" "How did the universe come into being?"—and even on what religious people call "God" (since for centuries, the concept of an all-encompassing unity was identified with God). We will also have to confront why monism is not more popular, if it follows so straightforwardly from quantum mechanics. Why does it sound so bizarre to us? Where does our intuitive, deprecative reflex come from? To really understand this bias, we have to venture into the history of monism.

The One is the story of both a serious crisis in physics and the half-forgotten concept that has the potential to resolve it. It explores the idea that "all is One," that matter, space, time, and mind are all just artifacts of our coarse-grained perspective onto the universe. Along the way it narrates how the concept evolved and shaped the course of history, from ancient times to modern physics. Not only did monism inspire the art of Botticelli, Mozart, and Goethe, but it also informed the science of Newton, Faraday, and Einstein. Even now, monism is becoming a tacit assumption underlying our most advanced theories about space and time. This is a story full of love and devotion, fear and violence—and cutting-edge science. In no small way, this is the story of how humanity became what it is.

1
THE HIDDEN ONE

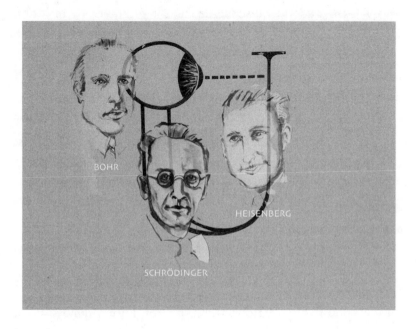

Q UANTUM MECHANICS IS THE SCIENCE BEHIND NUclear explosions, smart phones, and particle collisions. But it is more than that. It sketches a hidden reality beyond what we experience in our daily routines and holds within it the power to transform our notion of what is real—provided that it is taken seriously as a theory about nature. And therein lies the debate that begins our journey: How can we know that there exists something hidden that we can't experience directly? Doubts about this question launched the debate that ultimately returned the notion that "all is One" to the science most concerned with the separate identities and behaviors of the universe's most finicky bits and pieces.

Wheeler's U

"This guy sounds crazy. What people of your generation don't know is that he has always sounded crazy," Richard Feynman told Kip Thorne when the two had lunch together at an Armenian restaurant near the California Institute of Technology in 1971. Pointing at John Wheeler, their common PhD advisor sitting next to them, he went on, "But when I was his student, I discovered that, if you take one of his crazy ideas and you unwrap the layers of craziness one after another like lifting the layers off an onion, at the heart of the idea you will often find a powerful kernel of truth."[1]

John Archibald Wheeler belongs among the most influential physicists of the twentieth century. In his manners, his lifestyle, his political views, and his appearance, Wheeler was a rather conservative person. But in his ideas about physics, he displayed his wild side. This trait included a lifelong fascination with explosions, which almost cost him a finger when, as a young boy, he played with dynamite caps in his parents' vegetable garden. Working with Niels Bohr, the "grand old man" of modern physics, Wheeler later demonstrated that the atomic nuclei Uranium-235 and Plutonium-239 were possible candidates for nuclear fission and worked out how it might be accomplished.[2] That paper got published in 1939 on the day Adolf Hitler invaded Poland, and six years later those were the isotopes that fueled the explosion of the nuclear bombs deployed on Hiroshima and Nagasaki to terminate World War II.

When the Soviet Union tested its own first nuclear bomb in 1949, Wheeler and his students joined Edward Teller and Stan Ulam to become instrumental in the development and realization of the hydrogen bomb, taking advantage of nuclear fusion to fuel even more tremendous detonations. Wheeler's beloved brother Joe, fighting with the Allied forces against the German army in Italy's Po Valley, had been killed in action in October 1944 only weeks after Wheeler, who at that time was working on the development of the nuclear bomb, had received his postcard carrying only two words: "Hurry Up!" Ever after, Wheeler felt he had a "duty to apply [his] skills to the service of his country," as he explained in his memoirs.[3] Yet as much as he desired to "keep America strong," his heart was consumed by a deeper spirit of inquiry.[4] "From my earliest student days, I was most intrigued by questions about fundamentals. What are the basic laws that govern the physical world? How is the world, at the deepest level, put together? . . . What are the unifying themes? In short, what makes this world we live in tick?"[5] Wheeler loved to ask the deep questions: "How come the quantum?" "How come the universe?" "How come existence?"[6] "How come time?"[7]

His roughly fifty PhD students included superstars of physics, among them Richard Feynman, Kip Thorne, and Hugh Everett. Wheeler's discussions with Feynman paved the way for the quantum version of electrodynamics—providing a role model for any subfield of modern particle physics and earning Feynman the 1965 Nobel Prize. Together with Thorne and other students, Wheeler made Albert Einstein's theory of general relativity a respectable scientific topic again, culminating in the recent discovery of gravitational waves for which Thorne received the Nobel Prize in 2017. Wheeler also became the "grandfather" of the burgeoning field of quantum information[8]—the theory behind Google's, IBM's, Microsoft's, Intel's, and NASA's recent efforts to revolutionize computing—with his continuing interest in the foundations of quantum mechanics: the weird physics governing the microcosmos, for which Everett proposed his equally congenial and controversial interpretation suggesting the existence of many parallel realities, or "worlds." Finally, to top it all off, Wheeler is enshrined in the name of the Wheeler-DeWitt equation, the quantum equation for the wave function of the universe and starting point for much of Stephen Hawking's work on cosmology.

In addition to his accomplishments, Wheeler was famous for coming up with catchy phrases and names for new concepts. He popularized the name "black hole" for the timeless corpses of burned-out stars and coined the name "wormhole" for hypothetical, handle-like shortcuts between faraway regions in the universe. He deployed the

John Archibald Wheeler and his "U."

term "Planck scale" for the realms at tiny distances and extremely high energies where space and time themselves exhibit quantum properties and the name "quantum foam" for the bubbly consistency space and time supposedly have in these realms. And just as much as Wheeler loved catchy phrases, he loved to illustrate complicated concepts with simple sketches and diagrams.[9] Suitably, Wheeler's most enigmatic heritage is a little sketch depicting the history of the universe.

"There's the letter U. The U starts with this thin stem at the beginning when the universe is small. This stem gets fatter as we go up to the other side of the letter and at a certain point it's terminated by a big circle. And there is an eye sitting in there looking back to the first days of the universe," Wheeler said, describing his drawing that illustrates the evolution of the universe up to the emergence of conscious observers.[10] Indeed, as Wheeler emphasized, "We ourselves can get and do get radiation today from the early days of the universe," before he launched a bold speculation: "Insofar as the active observation has anything to do with what we ascribe reality to . . . then we can say this observer who was brought into existence by the universe has by his act of observation a part in bringing that universe itself into being."[11]

Was one of the most eminent physicists of the twentieth century seriously claiming that it is we who are responsible for the existence of the universe? That we ourselves, just by observing the world, create space and time and matter? And that this influence travels back in time to the beginning of everything and brings the universe into existence?

How can we follow Feynman's advice, unwrap the layers of craziness, and make sense out of "Wheeler's U"—assuming we discard the unsettling possibility that each time we look out the window, we unknowingly travel back to the beginning of time to initiate the Big Bang? After all, without constant acts of time travel, we obviously cannot really change the course of the early universe, to say nothing of actually creating it. The only possibility in which we "by our act of observation" can have "a part in bringing that universe itself into being" is by employing a radical reinterpretation: to adopt that what we experience as universe and as its history is only a specific perspective onto a more fundamental, hidden reality.

To make this point clear, consider a cylindrical object, such as a can of Coke. Depending on whether we view it from above or from

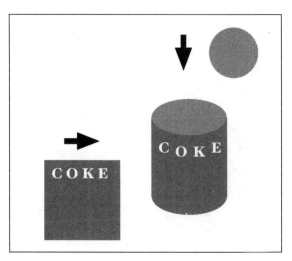

A can of Coke may look like a circle or a rectangle, depending on the perspective.

the side, the same can may be seen as a circle or as a rectangle. In this specific sense we can be held responsible for "creating" either the circle or the rectangle without really doing anything to the can itself, just by adopting a particular perspective.

We can obtain an even better understanding by comparing cosmic history with an old Hollywood movie. When we watch a film like *Bringing Up Baby*, the 1938 American screwball comedy starring Katharine Hepburn and Cary Grant, we experience a hilarious plot about a paleontologist trying to assemble the skeleton of a huge dinosaur, a project disrupted when he meets the crazy but beautiful Susan, who owns a tame leopard, and Susan's aunt's dog steals the last missing bone and buries it somewhere. But the story we experience in the theater is not really stored on the roll of film. Instead, a traditional movie projector displays the information on the film, one picture after another, flashed so quickly that the viewer has the impression of an unfolding story line. Again, the story is not really on tape; it is created by the viewer's perspective onto the projected film. The story is created by us watching it, while the original source of information remains unswayed, mounted on the projector. In the same way, cosmic history

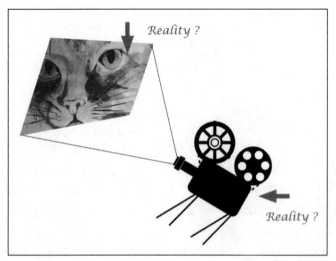

What is reality—the film roll mounted on the projector or the leopard story unfolding on-screen?

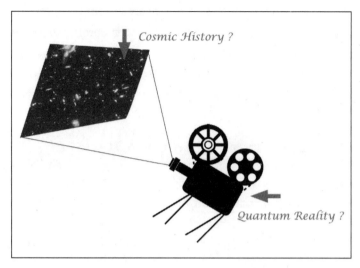

The Hollywood movie plot interpretation of cosmic history: Is the evolution of the universe just a projection of a more fundamental quantum reality?

(Credit: Shown on the screen is the Hubble Ultra Deep Field [HUDF]: original image by NASA, the European Space Agency, and S. Beckwith [Space Telescope Science Institute] and the HUDF Team, cropped and warped by the author.)

may be understood as what we experience, created by our perspective onto a fundamental "quantum reality."[12]

This "Hollywood movie plot interpretation of cosmic history" offers an astonishingly accurate picture of how quantum mechanics works, highlighting the most important question quantum mechanics forces us to ask: What is reality? Is it the light bulb and the collection of pictures stored on the film roll inside the projector, or is it the story experienced on-screen?

Even today, there are two camps of physicists and philosophers arguing fervently about exactly this question. The orthodox "Copenhagen" interpretation of quantum mechanics, advocated by Niels Bohr, Werner Heisenberg, and the overwhelming majority of physicists, insists that the movie plot constitutes reality. For many decades only a small number of outcasts, including (at least for some time) Erwin

Schrödinger, Wheeler's student Hugh Everett, and the German physicist H. Dieter Zeh, populated the "projector camp." This renegade view, however, is getting increasingly popular. It is part of a controversy that originated in the 1920s, as physicists fought out the question of how strange reality really is.

Mountaineering Through Thick Fog

Quantum physics grew out of an old argument about whether light and matter consist of particles or waves. In the first quarter of the twentieth century, a number of groundbreaking experiments had shown that both are the case. On one hand, the properties of electrons produced by light shining on material surfaces and the spectra of electromagnetic radiation emitted by heated objects could only be explained if light, so far considered to be a wave, consisted of indivisible energy portions, or "quanta." On the other hand, it had been demonstrated that the electron—so far known as a particle—also possessed the properties of a wave. In contrast to a particle, a wave has no accurately defined position; it is extended, or "nonlocal." If, for example, an electron is described by a wave, it exists at various different places at the same time until it is measured. In that instant the electron seems to collapse into a defined position that in general cannot be predicted accurately beforehand. Even worse, this puzzling behavior was only one aspect of a persistent problem: there was no consensus on what was more fundamental, particles or waves. Were particles just a secondary property of waves, or were waves just how particles behaved under certain circumstances? Or were both only imperfect projections of a deeper reality?

Who were the protagonists of these discussions, and why do the questions they raised occupy physicists even today? Above all, it was the ingenuity and work of four men, as well as their weaknesses, interactions, and personal relationships in the decade between 1925 and 1935, that gave rise to our understanding of quantum mechanics.

The Hidden One

Albert Einstein, born in 1879 and aged forty-six in 1925, was the most famous among the quartet.[13] With his maverick working style, Einstein was the archetype of the solitary genius. Working isolated from his colleagues as a clerk in a Swiss patent office, he published several groundbreaking works that included not only his theory of special relativity but also his hypothesis that light might consist of "quanta"—at least as a heuristic guess. While this idea earned Einstein the Nobel Prize in 1921, he never felt quite comfortable with the dual nature of light. For a time, Einstein almost completely abandoned his work on quantum physics to concentrate on generalizing his theory of relativity to include gravitation. In 1914, Einstein hadn't been able to resist leaving his beloved Switzerland and returning to his native Germany, where he had been offered a prestigious triple position, becoming the director of the newly founded Kaiser Wilhelm Institute for Physics, as well as a research professor at the University of Berlin and a member of the Prussian Academy of Science. Still, most of the time Einstein preferred to work alone.

Quite the contrary was Niels Bohr's approach to science. Born in Copenhagen in 1885, the thirty-nine-year-old Dane was a team player, having been a top-level goalkeeper in the squad of his brother Harald, who had won the silver medal with the Danish national team in the first Olympic soccer tournament in 1908. Niels Bohr loved to interact with a large group of young scientists being employed at or visiting his Institut for Teoretisk Fysik in Copenhagen, a place that soon developed into a Mecca for aspiring quantum physicists. Gathering the most brilliant young minds from all over the world around him, he became a mentor and fatherly figure not only to John Wheeler, who later adopted Bohr's style, but to a growing number of top-notch scientists who would move on to occupy professorships and research positions around the globe. While Einstein was concentrating on his theory of general relativity, Bohr had developed an imperfect yet useful model of the atom that resembled a miniature planetary system, with one notable exception: in Bohr's atom there existed only a restricted number of allowed orbits, which were later explained as standing waves by the French physicist Louis de Broglie. Assuming that electrons were hopping from one allowed orbit to another, without ever residing in the space in between, a process labeled by the physicists as "quantum jumping," Bohr's model allowed for the first time calculation of the characteristic frequencies at which hydrogen would absorb and emit light. It also secured Bohr the Nobel Prize in 1922.

Erwin Schrödinger, born in Vienna in 1887 and now thirty-seven years old, was a late bloomer and unconventional bon vivant with broad interests ranging from wine, theater, poetry, and art to Greek and Asian philosophy. Schrödinger's career had gotten interrupted when he was drafted to serve as an artillery officer in the Austrian army in World War I. Bored by the monotony of his daily routines and frustrated by the incapacity of his commanders, he immersed himself in physics books to keep sane. With his wife, Anny, whom he had married in 1920, he lived in an open relationship; both had affairs on the side. Schrödinger even kept a diary of his sexual encounters. Still, to Anny, he was a "racehorse" she wouldn't trade for a "canary."[14] After the war, Schrödinger had a string of brief employments in Jena, Stuttgart, and Breslau, until in 1921 he finally secured the professorship in Zürich previously held by Einstein. But then he fell ill: diagnosed with suspected tuberculosis, Schrödinger had to take a rest cure and stayed for nine months in the Swiss Alpine resort of Arosa. Thus, in 1925 the Austrian suffered from feelings of inferiority, unsure whether he would be able to leave a lasting mark in physics.

Werner Heisenberg, arguably the most ingenious among Bohr's protégés, was by far the youngest of the four men. Born in 1901, by 1925 the twenty-three-year-old Heisenberg had already earned himself a reputation as a physics prodigy. Since early childhood he had trained in solving mathematical puzzles and games, and he remained extremely ambitious. In his free time he loved to camp out and hike in the Bavarian mountains with friends from his Pathfinder group, the German variety of the Boy Scouts. Thus, when Heisenberg, now working as a postdoc at the University of Göttingen, struggled with the problem of the intractable electron orbits around the atomic nucleus in the spring of 1925, he compared this situation to an ascent in the Alps that he and his friends had undertaken the previous fall, during which they got lost in thick fog: "After some time we entered a totally confusing maze of rocks and pines . . . where we by no stretch of the imagination could find our path."[15]

A few months later, in May 1925, Heisenberg, plagued by hay fever, took sick leave to travel to the small island of Helgoland—a red rock devoid of bushes and meadows some forty nautical miles off the German coast in the North Sea. On Helgoland—inspired by the philosophy of positivism—he tried something new to tease out what happens inside the atom.

The credo of positivism holds that scientific theories should be based exclusively on what is observable in experiments. Scientists were urged to stick to what they saw in front of them, what they could measure and manipulate, rather than to theorize about an unobservable reality underlying the obvious phenomena. In other words, they were to concentrate on the on-screen reality and refrain from musing about the projector and film roll creating it. Accordingly, Heisenberg discarded the unruly electron orbits altogether. Viewed this way, the problem faced appeared vaguely reminiscent of the mathematical puzzles his father had assigned to him when he was a young boy. In his childhood, Heisenberg had excelled at these games, easily outperforming his elder brother. And indeed, Heisenberg now managed to find a solution where no one else had been able to before. In an ingenious act, Heisenberg developed an abstract formalism that allowed him, in a hard night's work, to calculate the energy levels of a simplified version of the atom, an oscillating spring. It was a "veritable calculation by magic," Einstein judged later.[16] Only a few months later, in early 1926, Heisenberg's friend Wolfgang Pauli adopted Heisenberg's formalism to calculate the energy levels of the hydrogen atom. Heisenberg and Pauli were enthusiastic: "Through the surface of atomic phenomena, I was looking at a strangely beautiful interior . . . nature had so generously spread out before me," Heisenberg wrote, describing his

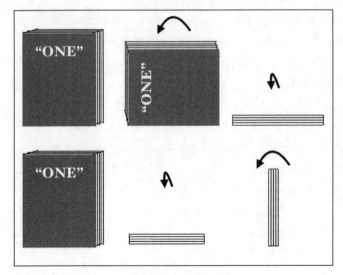

For matrices and rotations, the order matters.

exaltation;[17] Pauli rejoiced that he had found "a new hope, a new enjoyment of life."[18]

When Heisenberg returned to Göttingen, his supervisor, Max Born, soon realized that Heisenberg's strange algebra looked somewhat familiar: it fulfilled the multiplication rule of matrices, mathematical objects that can be used, for example, to describe rotations. In contrast to the multiplication of numbers, which can be performed in any order and still give the same result (like 2 × 3 = 3 × 2 = 6), order matters in multiplying matrices. For example, if you rotate the book in front of you first to the left and then toward yourself, the outcome will be different than if you performed the same operations in the reverse order. In Heisenberg's scheme, multiplication with matrices describes how probabilities for the properties of quanta evolve and how they can be observed. But in 1925, matrices were rather uncommon objects to most physicists. "I don't even know what a matrix is," Heisenberg had to confess at this stage.[19] In order to develop Heisenberg's scheme into a coherent theoretical framework, Born thus asked his twenty-two-year-old student Pascual Jordan, who, just like Born himself, had a solid background in mathematics, to join the project. Together with Heisenberg, over the following months they hammered out the first formulation of quantum mechanics, now known as "matrix mechanics."

Wunderkind Against Racehorse, Particles Versus Waves

Without a doubt, Heisenberg's glorious insight marked one of the greatest breakthrough moments in physics. But it also established the view that what happened inside the atom would elude any intuitive understanding. This became evident when Erwin Schrödinger in December 1925 found an equation that described the electron as a wave.

By way of backdrop, Schrödinger's marriage was in trouble. His wife, Anny, had had an affair with his best friend, the mathematician Hermann Weyl, while Weyl's wife had fallen in love with the physicist Paul Scherrer. This was too much even for the "racehorse" Schrödinger, who decided to hook up with an old girlfriend and leave Zürich to spend Christmas in Arosa. Only a couple of weeks before his trip, Schrödinger had become aware of de Broglie's hypothesis that electrons could be understood as waves. What was missing in this picture was an equation that described the energies and the time evolution of such a wave. During his two weeks in Arosa, Schrödinger must have experienced "a late, erotic outburst," Weyl imagined.[20] In fact, when Schrödinger returned to Zürich in early January, he carried with him the first sketch of an equation for quantum waves, convinced that "if I can only . . . solve it, it will be very beautiful."[21] With help from Weyl, by the end of January Schrödinger had not only solved his equation but also determined the hydrogen spectrum and submitted his results for publication.

Now there were two competing theories on the market. One described nature in terms of particles moving from one place to another via quantum jumps governed by probability rules, while the other described it through "deterministic," continuous waves. Once the state of Schrödinger's wave was known at one instant in time, its future evolution could easily be determined. In contrast to Heisenberg's matrix mechanics, Schrödinger's wave mechanics was elegant and intuitive, and mastering it didn't require mathematical tools the physicists weren't familiar with. Particles, Schrödinger concluded, would soon turn out to be nothing but a bunch of overlapping waves producing a lump of energy, similar to the occasional freak wave in the ocean.

Heisenberg was not convinced. "The more I think about the physical portion of the Schrödinger theory, the more repulsive I find it," he wrote to Pauli. "What Schrödinger writes about the visualizability of his theory is probably not quite right, in other words it's crap."[22] Even when Schrödinger had demonstrated that his approach reproduced the same results as Heisenberg's, the dispute raged on. Indeed, it turned out that Schrödinger's waves had problems. When interpreted as oscillating fields in normal space, they would dissolve too quickly to account for the particle-like behavior observed in experiments. Max Born showed that the wave's amplitude could be interpreted so as to provide the probability of finding a particle at the corresponding location,[23] and afterward he considered Schrödinger's quantum wave not as a real object but merely as a tool, "something purely mathematical," as Born described it.[24] By stipulating that the laws of quantum physics would yield only probabilities instead of concrete cause-and-effect relationships, Born sacrificed the principles of "causality" and "determinism" at the core of the old, "classical" physics since Isaac Newton: that nothing in the world of physics would happen without a cause and that knowing the exact state of a physical system at one time would make determining its future behavior possible. Bohr and Heisenberg agreed with Born. They appreciated that Schrödinger's formalism simplified many calculations but similarly dismissed the possibility that Schrödinger's waves had anything to do with the reality inside the atom. "Although Bohr was normally most considerate and friendly in his dealings with people, he now struck me as an almost remorseless fanatic, one who was not prepared to make the least concession or grant," Heisenberg remembered later.[25]

At this point, Albert Einstein felt increasingly uneasy. In the following spring of 1926, Heisenberg traveled to Berlin to give a colloquium. After his talk, Einstein invited the young man into his apartment. As soon as they arrived there, Einstein started to challenge Heisenberg's approach. Heisenberg would split the world into two separate realms: our daily-life, classical world (where objects have defined locations and properties and where causal physical laws determine their future) and a quantum realm that couldn't be described in everyday language. Even worse, Einstein criticized, since Heisenberg's formalism completely abandoned the notion of electron orbits inside the atom, it failed to elucidate the real nature of the quantum realm; it only summarized the observer's knowledge about the outcome of measurements. "You are moving on very thin ice," he warned Heisenberg.[26] Quantum mechanics had to be incomplete, Einstein felt. There had to be a hidden reality underlying the phenomena. Heisenberg left the meeting disappointed that he hadn't been able to convince the man he admired so much. Nevertheless, some of Einstein's arguments struck a nerve.

Right after meeting with Einstein, Heisenberg faced a tough choice. Having planned to accept another postdoc position with Niels Bohr in Copenhagen, the brilliant young man had also been offered a professorship in Leipzig. Less than three years earlier Heisenberg had almost failed his PhD exam, when he couldn't answer simple questions about the resolution of a microscope or a telescope or the functioning of a battery. Wilhelm Wien, the Nobel laureate of 1911 and head of experimental physics, had been frustrated about the young theorist's poor performance in his experimental lab course even before and was only grudgingly convinced by Heisenberg's advisor, Arnold Sommerfeld, to let the candidate pass, with a less-than-mediocre grade. Horrified, Heisenberg had literally fled Munich, taking the overnight train to Göttingen only to appear in front of Max Born the next morning, an embarrassed expression on his face and unsure whether he was still welcome to occupy his upcoming postdoc position. Now he was about to turn down the offer of a professorship in Leipzig, a tremendous honor for a scientist of his young age, in a time when hunger, poverty, and housing shortages were still common in postwar Germany. While his father, himself a professor of Byzantine studies, had urged him to accept the Leipzig position, Einstein and other senior physicists advised him to work with Bohr. Heisenberg decided to play for high stakes and went to Copenhagen. "I will always receive another call; otherwise I don't deserve it," he assured his parents.[27]

The stage was set for the development of the Copenhagen interpretation of quantum mechanics, a blessing and a curse for almost a century of research on the foundations of physics.

It Ain't What the Moon Did

By February 1927, Heisenberg's optimism was waning. Right after arriving at the Danish capital six months earlier, he and Bohr started their struggle to make sense out of quantum mechanics. While Heisenberg was perfectly happy with a mathematical formalism spitting out probabilities, Bohr insisted that physics should be framed in everyday language. As he detailed later, "By the word 'experiment' we refer to a situation where we can tell others what we have done . . . [T]herefore . . . the results of the observations must be expressed in unambiguous language"; he concluded, "All evidence must be expressed in classical terms."[28] Heisenberg pushed back: "When we get beyond this range of classical theory, we must realize that our words don't fit."[29]

The two men were at odds about another point too. While Heisenberg stuck exclusively to the idea of particles, Bohr wanted to incorporate Schrödinger's waves as well. Heisenberg had tried to discuss the matter with Schrödinger the previous summer, only to be chided again by Wilhelm Wien: "You must understand that we are now finished with all that nonsense about quantum jumps," the older man had told a dismayed Heisenberg before Schrödinger could even start to answer.[30] Now Bohr decided to invite Schrödinger to Copenhagen to discuss their inconsistent interpretations face-to-face. Schrödinger visited in September, fell ill, and was nursed by Bohr's wife, while Bohr sat on the edge of his bed, urging him to relent and accept that his theory was wrong. It didn't help; Schrödinger left without an agreement reached. In the next months, Heisenberg and Bohr continued their discussions day after day, often late into the night in an increasingly tense atmosphere. When, after many hours, both men found themselves almost in despair, Heisenberg would try to free his mind, strolling in the neighboring Faelled Park and asking himself again and again, "Can nature possibly be as absurd as it seem[s]?"[31]

Finally, Bohr decided he needed a break and set off on a four-week skiing holiday in Norway. Left behind, Heisenberg went on to ponder the problem of electron paths, only to hit "insurmountable obstacles" once again. "I began to wonder whether we might not have been asking the wrong sort of question all along," he remembered later.[32] Suddenly, Heisenberg recalled one of the arguments Einstein had raised against his first approach on quantum mechanics: "It is quite wrong to try founding a theory on observable magnitudes alone. In reality the very opposite happens. It is the theory which decides what we can observe."[33] Einstein's argument is known to philosophers as the "Duhem-Quine thesis": in order to extract an experimental result from an observation, it is necessary to understand what is happening during the measurement and how exactly the measurement apparatus and our perception function. "You must appreciate that observation is a very complicated process . . . Only theory, that is, knowledge of natural laws, enables us to deduce the underlying phenomena from our sense impressions," Heisenberg remembered Einstein arguing.[34] If theory determines what we can observe, Heisenberg thought, shouldn't it also determine what we can't observe?

Long after midnight, Heisenberg set off for another walk in the dark Faelled Park, and there he had the idea that would evolve into his famous uncertainty principle. Knowing the path of a particle would imply that one would know both the particle's location and its direction, its velocity at different instants of time. But when an experimentalist observes an electron in a cloud chamber, she doesn't observe the path itself but rather a sequence of localized interactions. As the water drops indicating the particle's position are much larger than the electron itself, this doesn't necessarily mean that both position and momentum (i.e., velocity times mass) are accurately known.

Checking this idea with his matrix formalism, Heisenberg discovered that it, in fact, wouldn't allow a simultaneous accurate determination of both position and momentum. In Heisenberg's version of quantum mechanics, matrices represented the measurement of observable quantities, such as position or momentum. The product of two matrices depends, however, on the order in which they are multiplied. This strange multiplication rule implied that it made a difference which quantity was measured first: determining a particle's position and after that its momentum would result in a different outcome than measuring in the reverse order. Now Heisenberg was able to represent what Wolfgang Pauli had described to him in a letter he had received in October: "One can see the world with the p-eye [i.e., momentum] and one can see the world with the q-eye [i.e., position], but if one opens both eyes together, then one goes astray."[35] As a consequence, the exact position and the precise momentum or velocity of a particle can't be measured at the same time—there always remains an uncertainty. Either the position remains unknown, or the momentum remains unknown, or both quantities are known only with limited precision.

Heisenberg felt vindicated: if a particle's position and velocity couldn't be pinpointed at the same time, it didn't make sense to talk about electron paths inside the atom. Either one doesn't know where the electron is, or one is ignorant about the direction in which the electron moves. With this insight, Heisenberg thought he had identified the origin of the breakdown of causality. "What is wrong in the . . . law of causality, 'when we know the present . . . , we can predict the future' is not the conclusion but the assumption," he wrote. "Even in principle we cannot know the present in all detail . . . [I]t follows that quantum mechanics establishes the final failure of causality."[36]

When Bohr returned from the ski slopes to Copenhagen, he brought only consternation. He promptly found a mistake in Heisenberg's argument and told him to rewrite the paper. At this point

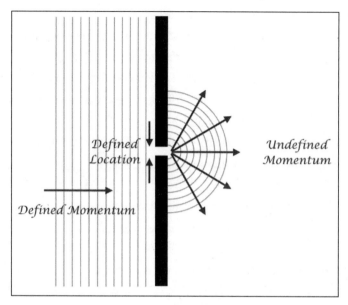

Illustration of Heisenberg's uncertainty principle: plain wave trains with a defined momentum spread out into all possible directions after hitting a barrier with a narrow hole in it.

Heisenberg literally broke down in tears.[37] What Bohr had realized was that Heisenberg's uncertainty, rather than invalidating the quantum wave picture, was in fact a typical behavior for waves. Long, plain wave trains have a well-defined momentum, but when these waves encounter a barrier with a narrow hole in it, they produce a circular wave behind it, spreading out in all possible directions. Limiting the wave to a narrow distance to determine its location thus results in a diverging momentum. Bohr thus identified the particle-wave duality as the core element of quantum mechanics and came up with an interpretation of his own: "complementarity."

Complementarity soon became the centerpiece of the Copenhagen grasp of quantum mechanics. But what exactly is complementarity? On a visit to Moscow, Bohr had scribbled the gist of his idea on the blackboard of his host's office: "Opposites [such as particles or waves] do not contradict but rather complement each other."[38] According to

Bohr both views, understanding matter as particles or as waves, had their justification, and each would reveal crucial information even though they seemed to contradict each other. As Bohr explained, "Evidence obtained under different experimental conditions cannot be comprehended within a single picture . . . [O]nly the totality of the phenomena exhausts the possible information about the objects," and "the study of the complementary phenomena demands mutually exclusive experimental arrangements."[39]

Heisenberg stubbornly refused to alter his paper, however. He had come to believe that his professional future now depended on a refutation of Schrödinger's wave mechanics, and he was convinced that he needed a fast publication to receive another job offer: "I have come to be in a fight for the matrices and against the waves," a "quarrel with Bohr," Heisenberg wrote to Pauli.[40] At this point the controversy between Heisenberg and Bohr escalated into a personal conflict. Finally, Heisenberg gave in and supplemented his uncertainty paper with a postscript, admitting that "recent investigations of Bohr have led . . . to an essential deepening and sharpening of the analysis . . . attempted in this work."[41]

This painful compromise between Heisenberg and Bohr became the centerpiece of what would be known later as the "Copenhagen interpretation," which would dominate both the thought of scientists and the textbook expositions future generations of physicists would learn from for at least the next fifty years. It would provide physicists with a working framework to approach the quantum mechanical problems they faced in atomic, nuclear, and solid-state physics, but it came with a price tag. In the Copenhagen philosophy, the act of measurement played a crucial role. According to Heisenberg, "Everything observed is a selection from a plenitude of possibilities," and only what was finally observed was considered "real."[42] This notion of a reality produced by the act of observation later fueled Wheeler's speculation about our creating the universe.

Coming back to the Hollywood movie plot interpretation of reality, the phenomenon of complementarity may be illustrated by a single film roll featuring different overlapping movies. Depending on the color of the light source or the angle of the projection, instead of *Bringing Up Baby*, the 1985 science fiction blockbuster *Back to the Future* might show up on the screen, to the bewilderment of a moviegoer ignorant of what kind of film roll is being loaded into the projector.

But as interesting as the concept is, the actual workings of complementarity remained somewhat vague. Bohr himself entertained at least two different versions of complementarity at different times. One described the relation of different projections or on-screen realities, such as particles versus plain wave trains with well-defined momentum. Another characterized the relation of on-screen realities and the underlying projector reality, the latter being described by Schrödinger's wave equation.[43]

At this point the physicists should have asked themselves what constituted these complementary observations. What kind of foundational reality underlay such conflicting experiences? In fact, whenever in the history of science physicists had discovered a more fundamental theory with a broader range of applications, they worked out how the old successful but limitedly valid theory could be understood as a limiting case of the new theory with its novel concept of reality. A famous example is Newtonian physics, which can be obtained as a low-energy limit of Einstein's special theory of relativity. But this was not done for quantum physics. In contrast to classical physics, quantum physics was capable of describing atomic and subatomic phenomena. Nevertheless, the Copenhagen physicists didn't understand classical physics as a limiting case of a more fundamental quantum or projector reality; rather they saw quantum mechanics as an instrument to obtain knowledge about classical objects experienced on-screen. The protagonists of the new quantum paradigm didn't take the plunge and explore the new reality hidden behind the quantum measurements. They left the quantum revolution unfinished. Instead, the Copenhagen interpretation evolved from compromise into dogma.

Naturally, Einstein wasn't happy. "The Bohr-Heisenberg tranquilizing philosophy—or religion?—is so delicately contrived that for the time being, it provides a gentle pillow for the true believer from which he cannot very easily be aroused," Einstein judged.[44] He maintained, "Does the moon exist only when you look at it? . . . I still believe in the possibility of a model of reality . . . that represents things themselves and not merely the probability of their occurrence."[45] Even more critical was the later assessment of H. Dieter Zeh: "This was an ingenious pragmatic strategy to avoid many problems, but, from there on, the search for a unique description of Nature was not allowed any more in microscopic physics . . . Only few dared to object that 'this emperor is naked.'"[46]

A World Split Apart

The next time Heisenberg and Einstein met was one and a half years later, at the Solvay Conference in Brussels. It was probably the most famous science meeting in history, with a participants' list that even today reads like a "who's who" of physics: Albert Einstein, Niels Bohr, Marie Curie, Max Born, William Bragg, Léon Brillouin, Arthur Compton, Louis de Broglie, Paul Dirac, Werner Heisenberg, Wolfgang Pauli, Max Planck, Erwin Schrödinger, and others. They gathered in October 1927 to discuss "electrons and photons" (photons being the specific quanta of the electromagnetic field) and the "new quantum mechanics." In Brussels, Bohr's and Einstein's views of the microcosm clashed, giving rise to the controversy that is ongoing even today. In subsequent years Bohr had to refute Einstein many times, one argument after another, typically expressed in the form of hypothetical thought, or *Gedanken*, experiments. But while Bohr succeeded case after case, he developed an interpretation of quantum mechanics that would become more and more absurd.

According to Bohr's and Heisenberg's Copenhagen interpretation, quantum mechanics was no longer a theory about nature. It was a theory about the experimentalist's knowledge about nature: a humanities concept rather than science. "One might be led to the presumption that behind the perceived statistical world there still hides a real world in which causality holds. . . . [S]uch speculations seem . . . fruitless and senseless. Physics ought to describe only the correlation of observations," Heisenberg argued.[47] Similarly, Bohr saw an "impossibility . . . of drawing any sharp separation" between the quantum object itself and its observation, "between . . . atomic objects and their interaction with the measuring instruments."[48] According to the Copenhagen physicists, atomic objects obtained their reality from the act of measurement. For Bohr, reality was like a movie shown without a film or projector creating it: "There is no quantum world," Bohr reportedly affirmed, suggesting an imaginary border between the realms of microscopic, "unreal" quantum physics and "real," macroscopic and classical objects—a boundary that has received serious blows by experiment since.[49] By installing this duality, Bohr enshrined what Einstein already had accused Heisenberg of doing when they debated in Berlin: Bohr had split the world apart.

For the fellow physicists in Brussels, however, Einstein's stubborn criticism arguing for an objective reality beyond what can be observed appeared increasingly as the obstinacy of an aging reactionary rather than as indication of a blind spot in the understanding of the foundations of physics. Einstein's friend Paul Ehrenfest captured the general impression of most physicists when he scolded, "Einstein, I'm ashamed of you, you are arguing against the new quantum theory just as your opponents argue about relativity theory."[50] "It was generally accepted by most physicists that Bohr won and Einstein lost," summarizes Leonard Susskind, one of the fathers of string theory and among the present day's most influential theoretical physicists. He continues, however, "My own feeling, I think shared by a growing number of physicists, is that this attitude does not do justice to Einstein's views."[51]

In hindsight, it is possible to identify where the discussions of Heisenberg, Schrödinger, and Bohr went astray. For the young Heisenberg, physics was a mathematical game that didn't necessarily reflect an underlying reality, thus there was no need for quantum waves. Both Schrödinger and Bohr, however, initially misunderstood quantum waves as objects in our everyday space. They didn't realize that, as the American philosopher David Albert accurately summarized, "Everything that has always struck everybody as strange about quantum mechanics can be explained by supposing that the concrete fundamental physical stuff of the world is floating around in something other, and larger, and different, than the familiar 3-dimensional space of our everyday experience."[52] Today we know, for example, that the quantum waves describing elementary particles such as neutrinos oscillate not between different locations but between the different types of neutrinos, a process known as "neutrino oscillations," the subject matter of the 2015 Nobel Prize. This abstract space of possibilities describes the quantum or projector reality in the Hollywood movie plot interpretation, and viewed this way, the struggle between Heisenberg and Schrödinger about particles versus waves boils down to the argument about whether a can of Coke is a circle or a rectangle. At some point, Bohr stepped in and judged that circle and rectangle are two complementary ways to experience the can, but we have to stick to the language of circles and rectangles and aren't allowed to talk about cans as cylinders.

The general acceptance of this interpretation became obvious when Heisenberg and Schrödinger were awarded Nobel Prizes in 1932 and 1933. Yet the fog remained. When Heisenberg teamed up with Bohr, they rather resorted to an interpretation that would define everything still covered in fog as "nonexistent." Later generations of physicists would refer to this airy philosophy as "fog from the north."[53]

Shadows on the Wall, Cats in the Box

The Hollywood movie plot interpretation of cosmic history that we have been using to understand quantum reality has a famous philosophical ancestor. In *Republic*, the ancient Greek philosopher Plato introduced an allegory describing a group of prisoners who live their entire lives in a cave where they are chained to a wall. All they ever see are the shadows of things created by a fire behind them. For these prisoners, the shadows on the wall seem to constitute reality. Then, one of the prisoners escapes and climbs up to the sunlight. He sees the real nature of things and realizes that everything he has known is a mere projection of this fundamental reality. When he returns to his fellow prisoners and tells them about the outside world, however, they don't believe him. They are too bound to their limited worldview to imagine an alternative.

In fact, this is more than just a vague metaphor for how quantum mechanics works. Remember that much of quantum mechanics' weirdness can be traced back to the fact that particles are described in terms of waves: just like wave trains stretching out over the surface of the ocean, objects can potentially exist in several places at once. If, for example, we have a situation where a particle can be located in two possible positions, such as "particle here" and "particle there," these positions correspond to single crests at the corresponding locations. In general, however, just as ocean waves can overlap and superimpose themselves on each other, before the measurement is performed, resultant waves corresponding to arbitrary superpositions of the two crests are possible. Even weirder, such an ambiguity isn't restricted to the location of objects but can be generalized to other properties as well.

In fact, any property of a quantum system could be described by a wave function, and just like waves describing the locations of particles, the corresponding properties can be superimposed. Just as swell spreads out over the entire surface of the ocean, so that surfers in Hawaii can catch waves originating from storms thousands of miles away, quantum waves exhaust all possibilities. Likewise, a quantum wave that hits a plate pierced by two slits will go through both of them. That's why, as far as the quantum wave is concerned, everything that can happen does happen. Only when the location or state of such particles is measured, a definite outcome gets observed, with a probability given by Born's stochastic interpretation of the wave function. Since the magnitude of the wave determines how likely it is to find the particle in one or another location, having one or another velocity or any other property, this suggests that before the measurement different realities ("particle here or particle there" or "fast versus slow particle") could coexist.

The monstrosity of this problem became obvious when, in 1935, Schrödinger generalized the situation to macroscopic dimensions. "One can even set up quite ridiculous cases," Schrödinger started and went on to devise a bizarre thought experiment: "A cat is penned up in a steel chamber, along with the following device (which must be secured against direct interference by the cat): in a Geiger counter, there is a tiny bit of radioactive substance, so small, that perhaps in the course of the hour one of the atoms decays, but also, with equal probability, perhaps none." While the radioactivity itself wouldn't harm the cat, a radioactive decay would trigger the release of poison into the cat's prison. "If [that] happens, the counter tube discharges and through a relay releases a hammer that shatters a small flask of hydrocyanic acid." Schrödinger's intention wasn't to torture cats but to illustrate the insane implications accidental microscopic processes could have for our everyday world. "If one has left this entire system to itself for an hour, one would say that the cat still lives if meanwhile no atom has decayed. The first atomic decay would have poisoned it. [One] . . . would express this [state] by having in it the living and dead cat (pardon the expression) mixed or smeared out in equal parts."[54] Schrödinger's cat illustrates a "macroscopic quantum superposition" and has become the classic example of quantum weirdness ever since.

In our daily lives, however, we never encounter any such quantum craziness. Objects have definite locations, and cats are either dead or alive and never in between. Upon observation, the potentialities represented by the quantum wave (or the film roll, as in the Hollywood movie example) then seem to "collapse" into a single, unique reality. It is this apparent "collapse" that corresponds to the projection of Plato's fundamental reality onto the prisoners' cave's wall. For Plato, the true reality was outside and not directly observable for the cave dwellers. Likewise, the orthodox Copenhagen interpretation was stuck in the perspective of Plato's prisoners—or the perspective of the cinema audience being ignorant of what happens inside the projector.

This point has been illuminated convincingly by John von Neumann, a Hungarian mathematician who played a leading role in the effort to understand quantum mechanics. In 1932, von Neumann published an influential textbook on the mathematical foundations of quantum mechanics in which he pointed out that, mathematically, a wave can be represented as a vector in a coordinate system. Two overlapping, superimposed waves then simply correspond to adding two vectors. Physicists and mathematicians call this construction a "Hilbert space," named after (Schrödinger's best friend) Hermann Weyl's advisor, the famous mathematician David Hilbert. In quantum mechanics, the axes of the coordinate systems are given by vectors corresponding to possible measurement outcomes, such as "particle here" versus "particle there" or "cat dead" versus "cat alive." But these states are not the only quantum states allowed. In such a coordinate system, vectors can be combined to produce superimposed realities: a vector, constructed by adding equal parts of the vector representing "particle

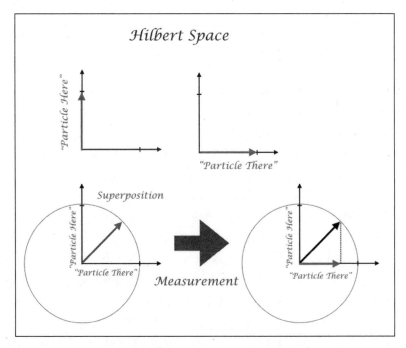

A quantum wave can be represented as a vector in Hilbert space. According to this picture, a measurement corresponds to a projection onto one of the axes.

here" or "cat alive" and of the vector representing "particle there" or "cat dead," corresponds to a quantum state giving equal 50 percent probabilities to finding the particle here or there or equal chances of observing a dead versus a live cat.

Thus, the quantum state before the measurement can correspond to any possible combination of the two axis vectors in a resultant vector of unit length. A particle being localized with equal chances of being "here" or "there" would be represented by a resultant vector pointing in the direction oriented at forty-five degrees with the coordinate axes. In general, all vectors on a circle around the vectors corresponding to "particle here" and "particle there" are allowed (or—more correctly—on a sphere, since the vectors can also be added with prefactors being complex numbers).

The measurement process is then described by what is known as the "projection postulate." The state of the quantum system before measurement is projected onto the coordinate axes, with a probability given by the squared component of the state vector parallel to the respective axis. In quite a literal sense, and just like in the Hollywood movie plot interpretation of cosmic history or Plato's cave, the quantum mechanical process of observation is understood as the projection of a more versatile prefiguration onto a concrete experience.

Von Neumann highlighted another important aspect as well. This projection process during a measurement was quite different from the continuous, deterministic evolution of the undisturbed state according to Schrödinger's wave equation. In contrast, the measurement corresponds to a sudden, nondeterministic jump into a classical state, now usually referred to as "collapse of the wave function." Thus, the measurement process is often described as "quantum-to-classical transition," and the difficulty of understanding the collapse of the wave function came to be known as "the measurement problem." In a lecture given at the University of Chicago in the spring of 1929, Heisenberg detailed that quantum mechanics could be considered either as a noncausal process "in terms of space and time" or as a causal process beyond space and time.[55] An obvious next step would have been to explore this causal description beyond space and time to find out what it entailed for the measurement problem and how classical reality, space, and time could emerge from the perspective of an observer. The fact that the quantum pioneers failed to do so has hampered the research for the foundations of physics ever since.

This is particularly baffling since von Neumann's representation of quantum systems as vectors in Hilbert space demonstrates that Bohr's separation of an unreal projector reality from the observable world is rather artificial. While it could make sense to assign the projector reality to a divine realm and see cinema operators as godlike creatures for the audience, following von Neumann, the on-screen reality resulting from a measurement is again a vector in Hilbert space that evolves according to Schrödinger's equation and can—just like the original projector reality itself—again be projected in another measurement. At least the vector representing the projector reality of a specific object is not special when compared to the collapsed object on-screen. What makes the projector reality special then? As we will see, it is its capability of merging several objects and even, in the extreme case, all objects in the universe into one.

* * *

Let's return to Wheeler's U—"a picture to inspire thought," as he himself described it, that is hard to make sense of.[56] Most of the time, Wheeler's U is understood as an illustration of his credo, "It from Bit":[57] the idea that matter originates from information, that "every particle, every field or force, even the spacetime continuum itself—derives its function, its meaning, its very existence entirely—even if in some contexts indirectly—from the apparatus-elicited answers to yes or no questions, binary choices, bits."[58]

But this interpretation leaves open the most puzzling aspect of Wheeler's drawing. If everything we experience is information, what is the "film roll," the "hardware," the fundamental fabric of the universe this information is stored on? For Bohr and Heisenberg, this hardware simply didn't exist.

Wheeler didn't give a decisive answer, and it is quite possible that even he himself didn't know what exactly he wanted to convey. As his former student and longtime collaborator Ken Ford, who worked with Wheeler on the hydrogen bomb, remembers, "I wouldn't say that . . . Wheeler literally believed in any of the ideas that he floated. Instead he hoped that they would inspire others—especially the next generation of physicists—to move the ideas from speculation to real physics."[59]

In a different place, though, Wheeler provided some clues: "The point is that the universe is a grand synthesis, putting itself together all the time as a whole . . . It is a totality."[60] He also speculated whether "a comprehensive view of the physical world [would] come not from the bottom up—from an endless tower of turtles standing one on the other—but from a grand pattern linking all of its parts."[61] The Hollywood movie plot interpretation can help to illustrate this point: On-screen, Susan, the paleontologist, and the leopard appear as distinct, individual characters. On the film roll, though, they are all mere features of a single camera shot.

Quantum mechanics goes even further. In quantum mechanics, so-called entangled systems get so completely and entirely merged that it is not possible to say anything at all about the properties of their constituents anymore. In quantum mechanics, all individual objects and all their properties result from the perspective of the observer—as, at least potentially, do matter, time, and space: they don't really exist on the film but are part of the story experienced as unfolding on-screen. In fact, this view again is strikingly similar to Plato's philosophy, which assumed that hidden on the most fundamental level there exists only one single object in the universe: the universe itself. Or, in the words of Plato, "The One."

2

HOW ALL IS ONE

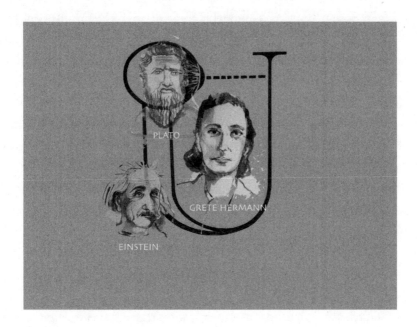

THE NOTION OF A HIDDEN QUANTUM REALITY BEYOND what we experience in our daily lives can be overwhelming. At the very least it triggers questions—such as whether we have any clue about what this hidden reality actually is. In what sense does the projector reality entail evidence for "monism," the belief in a hidden unity, or "One," behind all phenomena in nature? Remarkably, there is an operating mechanism that can merge reality into an all-encompassing unity. Enter quantum entanglement.

The Glue of Worlds

Quantum weirdness can be boiled down to three concepts utterly alien to our everyday experience: superposition, complementarity, and entanglement. Taken at face value, each of them marks a step away from classical physics as we know it. If superposition, the coexistence of alternative quantum realities, is weird, entanglement is even weirder. For Erwin Schrödinger, the man who came up with the mind-boggling, quantum mechanical zombie cat, entanglement constituted the core concept of quantum weirdness: "I would not call that one but rather the characteristic trait of quantum mechanics, the one that enforces its entire departure from classical lines of thought."[1] California Institute of Technology cosmologist Sean Carroll agrees, writing, "Without subsystems"—and thus entanglement—"Quantum Mechanics is trivial."[2] At the same time, according to quantum cosmologist Claus Kiefer, today entanglement is understood "as 'the' central element of quantum theory. Modern developments such as the field of quantum information are inconceivable without . . . entanglement."[3] Appropriately, when science writer Louisa Gilder had to choose a title for her vivid narrative about the history of quantum mechanics, she decided on *The Age of Entanglement*.[4]

In fact, entanglement is much more than just another weird quantum phenomenon. It is the acting principle behind both why quantum mechanics merges the world into one and why we experience this fundamental unity as many separate objects. At the same time, entanglement is the reason why we seem to live in a classical reality. It is—quite literally—the glue and creator of worlds.

Entanglement applies to objects comprising two or more components and describes what happens when the quantum principle that "everything that can happen actually happens" is applied to such composed objects. Accordingly, an entangled state is the superposition of all possible combinations that the components of a composed object can be in to produce the same overall result. It is again the wavy nature of the quantum domain that can help to illustrate how entanglement actually works.

Picture a perfectly calm, glassy sea on a windless day. Now ask yourself, How can such a plane be produced by overlaying two individual wave patterns? One possibility is that superimposing two completely flat surfaces results again in a completely level outcome. But another possibility that might produce a flat surface is if two identical wave patterns shifted by half an oscillation cycle were to be superimposed on one another, so that the wave crests of one pattern annihilate the wave troughs of the other one and vice versa. If we just observed the glassy ocean, regarding it as the result of two swells combined, there would be no way for us to find out about the patterns of the individual swells.

What sounds perfectly ordinary when we talk about waves has the most bizarre consequences when applied to competing realities. If your neighbor told you she had two cats, one live cat and a dead one, this would imply that either the first cat or the second one is dead and that the remaining cat, respectively, is alive—it would be a strange and morbid way of describing one's pets, and you may not know which one of them is the lucky one, but you would get the neighbor's drift. Not so in the quantum world. In quantum mechanics, the very same statement implies that the two cats are merged in a superposition of cases, including the first cat being alive and the second one dead and the first cat being dead while the second one lives, but also possibilities where both cats are half alive and half dead, or the first cat is one-third alive, while the second feline adds the missing two-thirds of life. In a quantum pair of cats, the fates and conditions of the individual animals get dissolved entirely in the state of the whole. Likewise, in a quantum universe, there are no individual objects. All that exists is merged into a single "One."

Einstein's Last Strike

When Albert Einstein and his younger colleagues Boris Podolsky and Nathan Rosen brought the quantum phenomenon of entanglement into the spotlight in May 1935, they didn't care about the merging of universes. They were focused on demolishing Niels Bohr's Copenhagen interpretation by deploying a newly devised weapon, now known as "the Einstein-Podolsky-Rosen (EPR) paradox."

By that time, Einstein had found himself a new home in America. When, two years earlier, the Nazis had seized power in Germany, Einstein had been on a visit at the California Institute of Technology and didn't "dare step on German soil" afterward.[5] On his return to Europe in the Belgian port of Antwerp, he asked to be driven to the German embassy in Brussels to hand over his passport and renounce his German citizenship. For the summer, he stayed in Belgium, while the Nazis enacted the law prohibiting Jews or anyone with a Jewish grandparent from holding an official position in German universities or academies. It was only one of many racist and inhuman measures that would culminate in the Shoah, the worst genocide in history. As a consequence, fourteen Nobel laureates and almost one-half of Germany's professors of theoretical physics were forced to emigrate. When Max Planck, as the president of the Kaiser Wilhelm Society, appealed to Hitler about the devastating effects this policy entailed for science, Hitler confirmed that "if the dismissal of Jewish scientists means the annihilation of the contemporary German science, then we shall do without science for a few years."[6] While Einstein still was in Belgium, newspapers reported that his name was on a list of assassination targets with a bounty of $5,000.[7] Rather than returning to Germany, in October 1933, he embarked an ocean liner for New York and joined a powerhouse of science on the rise: the newly founded Institute for Advanced Study in Princeton, New Jersey. It was from Princeton that Einstein launched his final attack on quantum physics.

Einstein had been unhappy with quantum mechanics ever since his discussion with Werner Heisenberg in the spring of 1926. One prominent feature of the theory Einstein didn't like was that according to quantum mechanics, nature on the most fundamental level appeared to be governed by chance; it was "probabilistic" rather than "deterministic." Einstein in contrast was convinced that "God doesn't play dice," a notion that would imply sacrificing the principle of causality.[8] But Einstein's conviction that "God doesn't play dice" wasn't his only problem with quantum mechanics. Einstein also couldn't accept the "nonlocality" of the theory, the fact that a particle with an accurately defined momentum wouldn't have a definable position, as a consequence of Heisenberg's uncertainty relation. Finally, probably Einstein's most severe point of concern was quantum theory's lack of an underlying reality: "At the heart of the problem is not so much the question of causality but the question of realism," he emphasized in 1950.[9]

In his 1935 paper with Podolsky and Rosen, Einstein combined these threads into a powerful argument that had the potential to undermine the comfortable trust in the Copenhagen version of quantum mechanics, or at least expose its bizarre consequences. The work came across "like a pike in a pond of carps," Schrödinger rejoiced.[10] The *New York Times* reported under the headline "Einstein Attacks Quantum Theory,"[11] while Bohr's assistant, the dismayed Leon Rosenfeld, commented, "This onslaught came down upon us as a bolt from the blue."[12]

Essentially, the three scientists considered the observation of an individual piece of a compound system. As a consequence of entanglement, the concrete conditions of the constituents can be totally unknown or undetermined. One can illustrate the gist of the argument in a fairly mundane setup. Think of two cans of paint being mixed together to make up the color green. By looking at the final result, you cannot infer whether the original colors were light blue and dark yellow or dark blue and light yellow or even dark green and white. Next, a chemist isolates the individual ingredients in her lab and carries one of them away without looking at it at first. Only after the chemist has brought home her container does she check the color of her ingredient. The crucial point is that as soon as anyone checks on one of the colors—whether it is light blue or dark yellow, dark green or white, the color in the remaining container is immediately fixed. But there is a crucial difference between mixing colors and quantum mechanics.

While the colors in both containers were already decided once the two containers got separated, even though they were not known yet, the state of a quantum system only gets determined when it is observed. When the experiment is performed, the wave function—according to Bohr's and Heisenberg's Copenhagen interpretation—collapses. Thus this collapse has to happen simultaneously everywhere in space-time. In other words, the condition of the observed component has to be transmitted at infinite speed, and, in particular, faster than the speed of light, to the other constituent—a phenomenon strictly forbidden by Einstein's special theory of relativity.

This pedagogic illustration is in fact similar to a later version of the EPR paradox devised by the American physicist David Bohm. Instead of colors, Bohm considered the spins of two particles that were produced, for example, in the same nuclear-decay process and then observed independently. In his version, the individual spins of the decay products add up to zero, implying they rotate in opposite directions, but only the measurement determines which particle spins left and which one spins right. The famous EPR work used yet a different example. It follows the same logic but nicely illustrates that in quantum mechanics, the properties of the components couldn't have been determined before they were observed. Just like Bohm, the authors considered two particles that were created in a decay process and then emitted back-to-back, but instead of spins, they discussed the positions and momenta of the particles. According to Heisenberg's uncertainty principle, an experimentalist could determine either the momentum or the position of one of the decay products but not both. If the experimentalist chose to measure the momentum of one of the particles, its location would be unknown and vice versa. On the other hand, since both particles originated from the same source, their momenta were necessarily related: if, for example, the decaying particle were at rest, the decay products would have equally large momenta pointing in opposite directions. Obviously, this implies that knowing one particle's momentum also determines the other particle's momentum. Correspondingly, if the experimentalist chose to measure the particle's position, also the position of its counterpart could be inferred. The problem with this reasoning is that, according to the Copenhagen interpretation, the first particle acquires a definite momentum only after an experimentalist has decided to measure it (and not to measure the particle's position instead, which would imply the particle had no well-defined momentum at all) and then actually performed the measurement. At the very moment when this measurement has been performed, however, the momentum of the second particle is fixed: a measurement of the first particle immediately collapses the wave function of the second particle. Since the same argument applies for a measurement of the particle's location, this implies either that both the momentum and the position of the second particle were set before the measurement (in contradiction of Heisenberg's uncertainty principle) or that one particle was somehow transmitting information about the measurement outcome to the other faster than the speed of light—without interacting with that second particle at all.

For Einstein, this was inconceivable. Not only would it contradict the core principle of his own special theory of relativity, according to which nothing can travel faster than light, but it also would constitute an effect without an interaction, dismissed by Einstein as "voodoo forces" and "spooky action at a distance."[13] The only explanation, according to Einstein and his coauthors, was that indeed both the position and the momentum of the second particle were determined before the measurement (as was the case in the example dealing with mixed colors), even if the quantum math wouldn't provide this information.

Quantum mechanics had to be incomplete, the authors inferred. According to the EPR paper's definition of reality, this information had to be provided by some principle beyond the formalism of quantum mechanics. Physicist Louis de Broglie, for example, had suggested theories amending quantum mechanics with so-called hidden variables that would specify the unobservable properties of quantum objects, an idea that later was adopted and further developed by David Bohm in 1952. Einstein rejected this solution as "too cheap."[14] Instead, he hoped for quantum mechanics to get completed within the framework of a more general, unified field theory, his own (and finally unsuccessful) pet project to develop a "theory of everything."

By taking advantage of entanglement, Einstein, Podolsky, and Rosen thus pointed out a mysterious connection among the constituents of composed systems. While the EPR paper wasn't the first treatment of entanglement ever—entanglement had been employed, for example, by the Norwegian physicist Egil A. Hylleraas in 1929 in his quantum mechanical discussion of the helium atom—it definitely brought the phenomenon into the focus of attention.[15] Moreover, Einstein and his coauthors exploited entanglement to highlight its paradoxical consequences for causality and locality, the cornerstone principles that every occurrence had an underlying cause and that causes could influence other events directly only at the place of their occurrence or later when being transmitted by a mediator with a velocity slower than the speed of light.

For Bohr, however, Einstein's work didn't really constitute a paradox. He replied with a series of papers that refuted Einstein's argument but at the same time led to an increasingly vague, incomprehensible, and flat-out absurd interpretation of quantum mechanics. Schrödinger wrote to Einstein that he found himself "snorting with rage" after reading Bohr's obscure, unsatisfactory response.[16] What for Einstein would imply that God was in need of "telepathic devices" was plausible for Bohr, who understood the quantum wave only as a description of knowledge rather than a description of reality.[17] Accordingly, for Bohr the collapse of the wave function wasn't a physical process changing the properties of the faraway particle but rather an update of the experimentalist's previous provisional knowledge. In a nutshell, Bohr denied that the properties of the second particle have reality at all before they had been measured, even if a wave function describing the second particle exists.

Many books have been written about this argument between Einstein and Bohr and what it implies for the notion of reality. Little has been said, however, about what this reality actually is. Following Einstein, who maintained that the only way to make sense out of this paradox was to conclude that quantum physics had to be incomplete, Bohm's version of the EPR paradox was later rephrased by John Stewart Bell and others in a way that allowed for experimental testing. Contrary to Einstein's and Bell's expectations, the predictions of quantum mechanics were confirmed. Yet such mysterious correlations between faraway objects are only one aspect of entanglement—and not the most interesting one. While today discussions of entanglement are often restricted to Einstein's "spooky action at a distance," three months after Einstein, Podolsky, and Rosen, Erwin Schrödinger published a paper that coined the term "entanglement" and clearly spoke about what the phenomenon actually means: "The best possible knowledge of a whole does not necessarily include the best possible knowledge of all its parts."[18] More explicitly, Schrödinger explained, "When two systems, of which we know the states by their respective representatives, enter into temporary physical interaction due to known forces between them, and when after a time of mutual influence the systems separate again, then they can no longer be described in the same way as before"—that is, "by endowing each of them with a representative of its own."[19]

Entanglement is quantum mechanics' way of integrating parts into a whole. Individual properties of constituents cease to exist for the benefit of a strongly correlated total system. Or, in the words of string theory pioneer Leonard Susskind, "It would not make sense for a mechanic to say, I know everything about your car but unfortunately I can't tell you anything about any of its parts. But . . . in quantum mechanics, one can know everything about a system and nothing about its individual parts."[20]

This insight entails consequences that are truly universal. After all, it makes sense to adopt the notion that all objects in the universe have interacted with each other at least somewhere along the way. If this weren't the case, these objects couldn't influence each other, and any assumption about the other's existence would be nothing but an unjustified hypothesis. Accordingly, entanglement shouldn't be confined to decay products or subatomic constituents. If interaction creates entanglement, this implies that the entire universe is entangled, as highlighted by Heisenberg's student and friend, physicist and philosopher Carl Friedrich von Weizsäcker, in his book *The Unity of Nature*: "The isolation of individual objects is always an approximation in quantum mechanics," Weizsäcker writes, eventually arriving at a radical conclusion: "If there could exist something which can be understood accurately as a quantum mechanical object, it would be the entire Universe."[21] As a consequence, "it seems necessary ... to give up the idea that the world can correctly be analyzed into distinct parts, and to replace it with the assumption that the entire universe is basically a single, indivisible unit," as David Bohm wrote in his 1951 textbook *Quantum Theory*.[22]

Entanglement provides the glue that allows quantum mechanics to constitute a monistic philosophy, the radical notion that there is but a single object comprising everything that exists—if quantum mechanics can be understood as a theory about nature and not, as the Copenhagen physicists maintained, a theory about knowledge. Einstein expressed the same thought in a more poetic way, when he wrote a letter of condolence to a grieving father whose son had died of polio a few days earlier: "A human being is part of a whole, called by us the 'Universe,' a part limited in time and space. He experiences himself, his thoughts and feelings, as something separated from the rest—a kind of optical delusion of his consciousness."[23]

The Greatest Idea

While such a conception appears far-fetched for the modern, rational mind, this apparently wasn't the case for our prehistoric or ancient ancestors. In fact, the notion that everything we experience boils down to different impressions of an eternal, unchanging primal reality isn't new. What may properly be characterized as "the greatest idea" is one of the oldest concepts known; it even might be as old as humanity itself. Monism, it seems, wasn't discovered or invented in a glorious moment of genius. As far as we know, it was always there.

We still can observe today that the idea of an all-encompassing unity is common across many indigenous religions in the Americas, Africa, Asia, or Oceania that often embrace a sacred or spiritual concept of nature. As the American evolutionary biologist and anthropologist Jared Diamond describes in his book *The World Until Yesterday*, a typical characteristic of peoples living in such traditional societies is "holistic" as opposed to "analytic" reasoning.[24] These small tribes and bands of hunter-gatherers or primitive herders and farmers were much more dependent on their natural and social environments and thus experienced the world as governed by interwoven networks rather than by the actions of individual agents. Instead of being characterized by innovation and progress, development and limited resources, their world was determined by natural cycles of life. Naturally, this experience got reflected in worldviews and belief systems. For example, to the American Indians of the Northeast, the Great Spirit "Manitou" inhabits—and thereby integrates or unifies—animals, plants, and inanimate objects such as rocks and can manifest in thunder or earthquakes.[25] In the traditional religion of the Hawaiian Islands, fondled by the constant trade winds, there exists the concept of "Ha," the breath of life, which gives rise to common Hawaiian words such as *aloha* (love, peace: in presence of the breath), *haole* (foreigner: person without breath), or *ohana* (family: those who share the breath).[26] More generally, such notions reflect belief in a "union of opposites and the relating of many strands in a lokahi, a harmony of diverse elements," as the Hawaiian philosopher of religion Gwen Griffith-Dickson writes.[27] A similar pervasive, vital force governing nature seems to exist in many African tribal religions.[28] In line with these observations, the British religious studies scholar Michael York includes "pantheism," the monistic belief that the universe is one and identical with god, as one of the main characteristic features of pagan theologies, next to "animism" (the belief that nature is animated), polytheism (the worship of many deities), and shamanism.[29] These findings suggest that monism originated from a simultaneous deification and unification of the universe prevalent in early societies, a hypothesis that may explain why monistic philosophies also in later times were often accompanied by an intensified appreciation of nature. In some societies, these holistic views of an interlinked world then evolved into full-fledged monistic philosophies.

Naturally, reliable evidence of how people envisaged the universe in ancient times has to be based on written testimony. The earliest coherent texts available date back to the third millennium BCE, after humans started to settle down in the Fertile Crescent, the "cradle of civilization" in the Middle East and neighboring regions, such as Sumer (in today's southern Iraq) and Egypt. Little known is that on February 16, 1923, almost exactly four years before Heisenberg discovered the uncertainty principle in Copenhagen, when the British archaeologist Howard Carter opened the tomb of Tutankhamun in the Egyptian Valley of the Kings, thunderstruck by the riches he found, he also encountered one of the earliest manifestations of monism. Next to mummies, the solid-gold coffin, the famous face mask, thrones, chariots, and more than five thousand other artifacts—"everywhere the glint of gold," as Carter described it[30]—there was a statue of "Neith," protecting the pharaoh's sarcophagus, as one of four tutelary deities. Neith, one of the oldest goddesses described in the Pyramid Texts, the ancient hieroglyphic writings carved into subterranean walls and sarcophagi starting from the third millennium BCE onward, was worshiped in the sanctuary of Sais, among the most important cult centers of Egypt, as "the mother and father of all things."[31] According to the ancient Roman author Plutarch, a veiled statue of the goddess in her temple bore the inscription "I am all that has been and is and shall be; and no mortal has ever lifted my mantle."[32] As Egyptologist Jan Assmann explains, "The Egyptians taught as a great mystery that God was all things, a spirit diffusing itself through the world and intimately pervading all things."[33] As startling as it is, as long as fifty centuries ago the ancient Egyptians knew something very similar to entanglement and actually adhered to the bold belief that everything that ever existed is amalgamated into a hidden One, a single, inaccessible being, symbolized by the veiled goddess Neith, later often identified with her better-known companion in protecting Tutankhamun's sarcophagus: the mother goddess Isis. According to this view, what we experience as nature is just the cover under which a hidden but looming unified fundamental reality can be surmised.

Some five hundred years after Tutankhamun had been laid to rest, and three thousand miles farther east, around 800 BCE, strikingly similar thoughts can be found in the Upanishads. In these ancient Sanskrit texts defining the spiritual core of Hinduism, the concept of "Brahma" is defined as holding together "all beings, all gods, all worlds, all breaths, all selves," just like "all the spokes are held together in the hub and rim of a chariot-wheel."[34] Like the ancient Egyptians with their goddesses Neith or Isis, the writers of the Upanishads knew an all-encompassing yet impersonal unity, signifying "the ground of the universe or the source of all existence, or that from which the universe has grown," as the Indian philosopher Telliyavaram Mahadevan explains.[35] In contrast, observable nature is understood as "maya," an artifice or illusion, portrayed in the Upanishads as follows: "Like a conjuring trick, it is made of maya. Like a dream, it is a wrong seeing... Like a wall painting, it delights the mind, but deceptively."[36] Or a veil, as nineteenth-century philosopher Arthur Schopenhauer emphasized, "the veil of deception that covers the eyes of mortals and lets them see a world that cannot be described as either being or not being: for it is like a dream; like sunlight reflected off sand that a distant traveller mistakes for water."[37]

Another two hundred years later and twenty-six hundred miles farther east, the sixth-century-BCE Chinese sage Lao-tzu defined in his book *Tao Te Ching* the concept of "Tao" or "Do" as "the beginning of heaven and earth" and "the ancestor of the myriad creatures."[38] Literally meaning "path" or "way," Tao is, in fact, yet another name for "The One": "Understood in this way, we can see that it is 'the One' or the 'tao' which is responsible for creating as well as supporting the universe," explains Darrell Lau, professor of Chinese at the University of London.[39]

These are only a few of many examples. Monistic concepts proliferate across philosophies and religions, in Mahayana and Zen Buddhism,

Christian mysticism, Islamic Sufism, and Sikhism. They have been claimed to be a—or even "the"—core concept of a "perennial philosophy," a hypothetical, metaphysical truth shared allegedly by all religious traditions as advocated, for example, by the English writer Aldous Huxley.[40] In fact, while there exists solid evidence that this philosophy spread out from specific sources and regions, such as ancient Egypt, it is not unlikely that monism emerged in various cultures and geographic regions independently, that it constitutes a "universal primal concept," as advocated by the German philosopher Karl Albert.[41] At least one can safely conclude from the global presence of monistic philosophies that an all-encompassing "One" exerts a universal fascination.

Such ancient testimonies are usually understood as mythological, with little relevance for modern science, if any at all. Yet, on a closer look, they address the very same problem that occupied John Wheeler: "How is the world, at the deepest level, put together?" What is the ultimate reality? On what foundation should the scientific endeavor be based? Just like the on-screen and film-roll realities encountered in the discussion of quantum mechanics, the ancient myths distinguished between an experienced reality, described as "illusion," "veil," or "maya," and a fundamental inaccessible reality referred to as "Brahma," "Tao," or "One." This striking parallel was emphasized by the French physicist Bernard d'Espagnat when he titled his 1995 textbook on quantum mechanics *Veiled Reality*—a probably deliberate reference to the Hindu concept of maya or to the veiled Egyptian goddess in Sais.[42] "If quantum theory appears as a 'smokey dragon,' the dragon itself may now be recognized as a universal wave function, partially veiled to us local beings by the 'smoke' of its inherent entanglement," H. Dieter Zeh explained a few years later.[43] Yet the idea of such a hidden fundamental reality doesn't exhaust the similarities of modern physics and ancient mythology. Just like quantum mechanics, the monistic philosophies knew a concept of complementarity: the fundamental Brahma, Tao, or One was a conjunction of opposites, of complementary features experienced in the daily-life, on-screen reality—just like the fundamental film-roll or quantum reality could manifest itself in different projections, such as particles or waves. Finally and most strikingly, modern science and ancient mythologies even seemed to arrive at a similar conclusion about what this fundamental reality is: "it is a grand synthesis, putting itself together all the time as a whole . . . It is a totality."

The Aborted Revolution

With quantum physics, Heisenberg and Bohr had discovered an entirely new realm of nature, a quantum reality underlying and unifying everything in the universe and obeying strange new laws of physics. Yet, instead of setting out to explore this uncharted territory, they decided to declare it nonexistent. The revolution initiated by the groundbreaking discoveries of Heisenberg, Schrödinger, and many others wasn't completed but rather aborted just when it was about to give away the foundations of physics. Yes, the objects populating this realm, the film-roll or projector reality behind both particles and waves, wasn't directly observable, and Heisenberg and Bohr were influenced by the philosophy of positivism, which considered only those things "real" that were accessible by experiment. Yet neither Bohr nor Heisenberg was a devoted, hard-core positivist. For example, Heisenberg wrote in his autobiography, "The positivists have a simple solution: the world must be divided into that which can be said clearly and the rest, which we had better pass over in silence," only to dismiss this conviction: "But can one conceive of a more pointless philosophy, seeing that what we can say clearly amounts to next to nothing?"[44]

So why then did Heisenberg and Bohr stop at the frontier of the quantum realm and not press on? Why does the Copenhagen interpretation insistently "undervalue the role that the mathematical structure of an empirically successful theory can play in accessing the modal, physical, and metaphysical nature of the universe," as the American philosopher Nora Berenstain determines?[45] Where did this self-restriction of the Copenhagen physicists come from? And how did it transform into a dogma?

It wasn't so much that the early quantum luminaries were totally ignorant of the monistic implications of quantum mechanics. In his essay "Physical Science and the Study of Religions," Bohr, for example, wrote that "the essential wholeness of a quantum phenomenon finds its logical expression in the circumstance that any attempt at its subdivision would demand a change in the experimental arrangement incompatible with its appearance."[46] Bohr was also well aware of the parallels between complementarity and the conjunction of opposites speculated about in ancient monistic philosophies. When in 1947 Danish king Frederick IX announced that he was conferring the Order of the Elephant, Denmark's highest honor, on Bohr, Bohr designed his own coat of arms, which featured a yin-yang, the pictorial representation of Lao-tzu's Taoist philosophy that seemingly opposite forces in nature are actually complementary on a deeper level of understanding. As a motto, Bohr added in Latin *Contraria sunt complementa* (opposites are complementary). In a similar spirit, Heisenberg titled his autobiography *The Part and the Whole*.[47] When Fritjof Capra interviewed Heisenberg in 1972 for his book *The Tao of Physics* and inquired about the famous physicist's thoughts on Eastern philosophy, Heisenberg told Capra, to his great surprise, "not only that he had been well aware of the parallels between quantum physics and Eastern thought, but also that his own scientific work had been influenced, at least at the subconscious level, by Indian philosophy."[48]

This is among the reasons that it is so baffling that Bohr and Heisenberg wouldn't want to scrutinize where our on-screen daily-life reality originates from. On closer inspection it appears quite likely that the motivation of the Copenhagen physicists for discarding the quantum realm as "unreal" was triggered not exclusively by the fact that it wasn't observable and at least as much by what it was: an all-encompassing unity, a concept that historically had been associated with religion and often had been identified with God. Even more so, it was one of the core concerns of Christian theology for two thousand years to make absolutely sure that such a concept wouldn't be considered part of nature, that there existed a sharp distinction between God and world.

Obediently, the Copenhagen physicists pushed the reality behind our immediate observations into the sphere of religion.

In fact, it wasn't only Einstein who understood Bohr's interpretation of quantum mechanics as a "tranquilizing philosophy or religion." In this context it is revealing that in many conversations in which Heisenberg recollects discussing with Bohr the deeper meaning of complementarity, the conversation soon digressed into an argument about the relationship of science and religion. As Heisenberg remembers, Bohr—for whom language was essential for what could be spoken of in the first place and for whom the complementarity of different on-screen realities such as particles versus waves betrayed the limitations of what we can say—argued that "we ought to remember that religion uses language in a quite different way from science."[49] In this context it is important to observe that for Bohr, complementarity wasn't only a "horizontal" relationship, describing the link between different on-screen projections such as particles and waves, but applied also to the "vertical" relation of on-screen and film-roll or projector realities. Just like particles and waves were mutually exclusive but equally justified descriptions of nature, so were experiment-driven science and metaphysics, Bohr believed. In his autobiography, Heisenberg recounts discussions about the schizophrenic conviction that religion and science refer to quite distinct facets of reality: "This view which I know so well from my parents associates the two realms with the objective and subjective aspects of the world."[50] While Heisenberg confesses that "he doesn't feel altogether happy about this separation," he nevertheless credits this philosophy with having pacified the centuries-old conflict between science and religion.[51]

Likewise, Heisenberg's friend Wolfgang Pauli considered rational scientific thinking and irrational mystical experience as "complementary" approaches to insight,[52] and Heisenberg's collaborator Pascual Jordan argued that the indeterminism in the quantum measurement process would invalidate the "conviction that there exists a coherent causal naturalistic world, that nature . . . wouldn't allow for an intervention of a divine creator . . . that the de-enchantment of the world by the sciences would be an inevitable result of scientific research."[53] For the Copenhagen physicists, the reality behind particles and waves was none of their business. It was the "tao that can't be named," the things in heaven above. "Thou shalt not make unto thee any graven image, or any likeness," as the Second Commandment demands.[54] Instead of realizing that the monism embraced in ancient philosophies is indeed a crucial concept for modern physics, the Copenhagen physicists reclassified the foundation of physics as religion.

By installing the dogma that "there is no quantum world," the Copenhagen physicists revived a narrative that originated among Christians in late antiquity and the Middle Ages and that claimed a strict separation between the material world and a divine realm. At the same time, they assigned the nature of quantum objects to this divine realm. In the years to come, when quantum mechanics was applied successfully as a working paradigm in nuclear, particle, and solid-state physics, the physicists embraced a pragmatic attitude toward its philosophical foundation. Richard Feynman, for example, recommended, "Do not keep asking yourself . . . but how can it be like that. Nobody knows how it can be like that"[55]—an attitude fittingly summarized by the American physicist David Mermin as "Shut up and calculate!"[56] Musings about the meaning of quantum mechanics were generally considered a private pastime, only loosely related to and not quite significant for doing real physics.

So if it wasn't a consequence of ignorance, how come Bohr's and his followers' philosophy of denial? As Aage Petersen realized, "Bohr's philosophical ideas were not originally inspired by physics, but the characteristics of the new theory fitted his philosophy wonderfully well."[57] Thus the history of quantum mechanics is only one part of the story. The history of monism and its troublesome relation with religion is another.

The Woman Who Wouldn't Obey Authority

Beyond any doubt, Einstein, Bohr, Heisenberg, Schrödinger, and many other of the early quantum pioneers have to be ranked among the greatest scientists of all times. If even these luminaries failed to obtain a truthful comprehension of what the theory really means, one may reasonably wonder whether it was possible at all. In other and better times, maybe Einstein could have found an ally in the young mathematician and philosopher Grete Hermann. Hermann was an early critic of the Copenhagen gospel who anticipated much of what would later become integral parts of a more justified and coherent interpretation of the reality quantum mechanics entails. But she was disadvantaged as a woman, a scientific outsider, and a conscientious person in savage times.

Growing up as one of seven children of a sailor and merchant in Bremen, Hermann had studied mathematics and philosophy in Göttingen, receiving her PhD for work under the supervision of Emmy Noether, the genius scientist who pioneered the search for symmetries and conservation laws in physics and, after a long struggle, had finally become the first woman to be allowed to teach mathematics at German universities. Following her PhD Hermann began working with the philosopher Leonard Nelson, a fervent advocate of Kantian philosophy and democratic socialism. In the 1930s Hermann got interested in quantum mechanics, after she learned that Heisenberg had claimed repeatedly that "the meaninglessness of the causal law is definitely proved," for example in the *Berliner Tageblatt*, by then one of Germany's high-circulation daily newspapers.[58] To Hermann, for whom the law of causality was a condition for empirical research rather than its result, such a claim was nonsensical: "The causal law is not an empirical assertion which can be proved or disproved by experience but the very basis of all experience."[59] Unafraid of any authorities, she "decided to fight the matter out" directly with Heisenberg.[60] In the spring of 1934, Hermann moved to Leipzig to discuss the issue in Heisenberg's seminar. This was everything but an easy task, and the bold mathematician had to struggle to be taken seriously in the close-knit, exclusively male group around Heisenberg, Friedrich Hund, and their young protégé Carl Friedrich von Weizsäcker:

> *Quite often I encountered a lack of understanding and impatience . . . when I asked for the physics reasons behind the limits of determinism and whether they could be circumvented by so far undiscovered hidden parameters. Only Heisenberg took this question seriously that, as I only learned much later, had also been discussed fervently in the famous Einstein-Bohr discussions. I still see Fritz Hund's friendly ironic smile before me, when he answered my challenge by asking whether I would believe the limits of determinism could be circumvented by the discovery that electrons had more or less red noses.*[61]

Just like Einstein, Hermann initially believed quantum mechanics was incomplete and needed to be supplemented. If quantum mechanics provided only probabilities for what would be observed in an experiment, shouldn't it be possible to amend the theory with additional information about the exact state of the physical system so that it would provide an accurate prediction? The physicists, however, felt confident that it wasn't, reassured by the authority of John von Neumann.

Like Hermann, von Neumann was the product of Göttingen's school of mathematicians around David Hilbert. Early on the Hungarian mathematician had earned himself a reputation as a prodigy and universal genius. Von Neumann had started to work on logics before he focused his attention on quantum mechanics. Later he contributed crucially to the design of nuclear weapons, revolutionized economics with the development of game theory, and devised the architecture of the modern computer. In his influential 1932 book *Mathematical Foundations of Quantum Mechanics*, von Neumann had developed and summarized the mathematical groundwork of the new theory. In this context he also presented a proof that it would be impossible to supplement the theory with the so-called hidden variables: parameters that would specify the state of a quantum object before measurement. Most physicists now agree that hidden variable theories are indeed most likely a dead end since they are difficult to reconcile with special relativity. Nevertheless, they constitute a valid and interesting possibility. In her struggle to vindicate causality, Grete Hermann started out by demonstrating that von Neumann's proof was wrong: as Hermann pointed out, von Neumann's argument relied on an unjustified proposition that was correct for quantum mechanics but not necessarily for a quantum theory with hidden variables. While this was a remarkable achievement, nobody took notice. Only twenty years earlier Hermann's genius supervisor Emmy Noether had been denied a concession to teach at universities—since she was a woman. Now a woman philosopher came and argued she had proven wrong the entire physics community, including the famous von Neumann. Even scientists are not immune to bias and prejudice.

Making matters worse, Hermann's work was published in a rather obscure journal and in turbulent times. Within weeks of seizing power in January 1933, Hitler's Nazi Party had started to dismantle the fragile German democracy, to establish a brutal and racist regime, and to displace, torture, and murder political opponents and Jews, a bitter foretaste of looming genocides. In such times, physics and philosophy weren't the most important activities in Hermann's life. How different Hermann's priorities were, as compared to those of the physicists she was in discussion with, is evident from a revealing exchange of letters with Heisenberg in the summer of 1937, four and a half years after the Nazi terror began. When Heisenberg wrote that he wouldn't be present in Leipzig in September since he had been drafted for a military exercise, confessing that he was actually "looking forward to it, being forced this way to a rigorous change of his daily routine," Hermann was appalled: "Your joyful anticipation . . . has evoked in me a fatal emotion. Not of course since you appreciate frugality and toughening up, but since you at the same time for yourself and others embrace the authorities that impose this change of routine upon you. I can't believe that this appreciation results from a conscious and thoughtful approval of this authority's objectives."[62] Outspoken as ever, Hermann rebuked the famous Nobel Prize winner: "You allow these authorities to force persons, in this case also yourself, by explicit or implicit force how they have to live their lives, instead of living according to their own conviction"—a freedom, Hermann emphasizes, that she considers crucial for "one's life having a meaning at all."[63] Hermann's courage is remarkable, even more so since her letter could easily have put her life in jeopardy had the authorities somehow become aware of it. The naivety of Heisenberg's reply is distressing: "I'm just thankful that the society gives me an opportunity to participate, without having to change my convictions. To scrutinize the political justification of the institution—this would only make sense anyway if one would assign oneself the task to change the world politically, which I believe would be possible only as an alternative if one would stop doing science."[64]

In contrast to the politically uninterested Heisenberg—his wife titled her own narrative about her husband's biography "the political life of an unpolitical person"[65]—Hermann wanted to be both a scientist and a political person, especially in a time of injustice and terror. While Hermann worked as a daytime mathematician and philosopher, at night she turned into an underground activist against the Nazi regime. Her resistance group, founded by her supervisor Nelson, published an "urgent appeal" to all opponents of fascism to join forces, warning that "the eradication of all personal and political freedom in Germany is imminent."[66] It was signed by Einstein and famous artists and authors. The group also tried to build a clandestine labor union and sabotaged the inauguration ceremony of the new "autobahn" highway by installing anti-Nazi slogans on all bridges, which later had to be ploddingly cut out of the Nazi propaganda videos. It was an extremely courageous and perilous activity; only a few years later, for example, the Munich students who formed the resistance group known as "White Rose" were sentenced and put to death for far less.

Soon after her discussions with Heisenberg's group in Leipzig, the pressure of persecution in Germany became too large, and Hermann had to emigrate to Denmark. From there, Hermann published two papers in 1935, both titled "The Foundations of Quantum Mechanics in the Philosophy of Nature." In the second of these papers, Hermann changed her focus. Instead of trying to amend quantum mechanics with hidden variables to restore the determinism of quantum states in the on-screen reality, she concentrated on the film-roll reality: "The theory of quantum mechanics forces us . . . to drop the assumption of the absolute character of knowledge about nature, and to deal with the principle of causality independently of this assumption."[67] In a move that would never have occurred to Bohr, she decided to take the film-roll reality seriously and even adopted it as "nature," as more fundamental than the daily-life, on-screen reality. In doing so, Hermann made an important observation. She realized that the apparent causality violation experienced in quantum processes was nothing but an artifact of the on-screen reality: "Quantum mechanics has therefore not contradicted the law of causality at all, but has clarified it and has removed from it other principles which are not necessarily connected to it."[68] As von Neumann had already emphasized in his famous book, quantum mechanics consisted of two processes: first, the deterministic smooth evolution of quantum waves according to Schrödinger's equation, and then the sudden indeterministic collapse during the measurement collapse. Now Hermann argued that quantum mechanics in fact provides the causal reasons for the state of an observed system, albeit only in relation to the specific observation, and that the nondeterministic feature was a property of the projection, not of the film roll. But if quantum observations are relative to the specific outcome of the measurement, and if there are many such possible outcomes, this implies that there are many possible observations and observers. As pointed out by Dirk Lumma in the introduction to his translation of Hermann's work, twenty years later quite similar thoughts would lead to a new perspective on the notion of reality in quantum mechanics.[69] It was named "relative state formulation" but it became famous under a different name: the "many worlds interpretation."

Rather than confining the formalism of quantum mechanics to our everyday prejudice, as attempted in hidden variable theories, the many worlds interpretation would expand the realm of reality. Both hidden variable theories and the many worlds interpretation take quantum mechanics seriously as a theory about nature. Hidden variable theory essentially strives to confine the potentialities of quantum mechanics to comply with our perceived notion of reality. The many worlds interpretation instead widens our notion of reality to comply with what quantum mechanics predicts. But to interpret quantum mechanics this way, one would need to understand why we experience many things if the world is one after all. Hermann's insights prove that such a grasp of what quantum mechanics entails indeed was possible already in the 1930s. Rather than the implications of the mathematical formalism, it was philosophical bias (most notably Bohr's), personal motives (such as the insecurities and competition of Heisenberg and Schrödinger), and historical coincidences that gave rise to the Copenhagen interpretation. In the following years, however, both the community of scientists in general and Hermann's life in particular were drawn steadily into the maelstrom of the global catastrophe.

* * *

Soon most physicists who had been wondering about the strangeness of quantum mechanics before were busy developing nuclear weapons and other militarily relevant technology. Hermann's mentor Nelson had died in 1927 at the early age of forty-five. In 1935, her advisor Emmy Noether, who was Jewish and had to emigrate to the United States, also died from complications after a pelvic surgery to remove a tumor. Soon after, Hermann herself fled to England and entered into a marriage of convenience to avoid getting interned as an alleged enemy alien, so that she could continue her political fight. After the war, she returned to Germany, where she struggled to rebuild a democratic society, the Social Democratic Party, and a modern curriculum for teachers.

Hermann's criticism of the quantum gospel remained unnoticed until more than thirty years later, when John Stewart Bell independently discovered the weak point in von Neumann's proof. While back in 1934 Grete Hermann was discussing with Heisenberg and Weizsäcker, Einstein already had settled at the Institute for Advanced Study in Princeton, as had Hermann Weyl, John von Neumann, and his lifelong friend Eugene Wigner, the physics Nobel Prize awardee of 1963. And in 1938, John Wheeler got hired by the neighboring Princeton University. Twenty years later, a PhD student of Wheeler's, a certain Hugh Everett III, would again dare to take quantum mechanics and its implications for the film-roll reality of nature seriously and argue, "Let's just believe the basic equations—what's this extra jazz for?"[70]

3

HOW ONE IS ALL

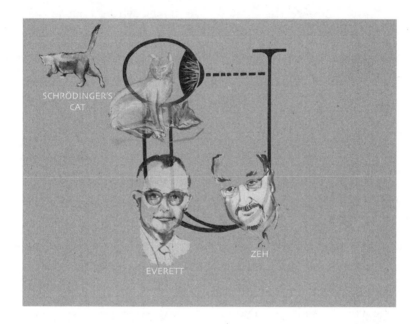

I F "ALL IS ONE," WHY DO WE EXPERIENCE THE WORLD AS plurality? Where do matter and structure originate, and what actually happens during a quantum measurement? As we will see, the entanglement of observer and observed and the quantum phenomenon of decoherence help explain how a hidden One is experienced as emergent individual objects in parallel realities and worlds. Yet, whenever progress toward this literal interpretation of quantum mechanics was made, it was greeted with fierce opposition and hostility by the physics establishment. Two physicists, a prankish nerd and a sober bulldog, ultimately achieved the breakthrough to explain how quantum mechanics, the fundamental reality, and our everyday experience were related. These remarkably dissimilar characters arrived at essentially the same conclusions about how the universe functions.

The Player and His Game

"A new scientific truth does not triumph by convincing its opponents and making them see the light," Max Planck wrote, "but rather because its opponents eventually die."[1] And irrespective of whether the Copenhagen interpretation was truth or not, its most prominent critic didn't live forever. On April 14, 1954, Albert Einstein gave his last seminar, reinforcing his criticism of quantum mechanics as a guest speaker in John Wheeler's relativity class. One young man, eagerly listening in the audience, later remembered, "[Einstein] put his feeling colorfully by stating he could not believe that a mouse could bring about drastic changes in the universe simply by looking at it."[2]

When, a few months thereafter, the young man signed on to become a PhD student of Wheeler's, his new advisor soon realized that this student was special: "independent, intense, driven"[3] and "highly original."[4] Less than three years later, the "original young man" would boldly characterize Einstein's EPR paradox as "fictitious."[5] At the same time, he would dismiss John von Neumann's wave function collapse as "untenable" and describe the Copenhagen interpretation as "overcautious," "hopelessly incomplete," and "a philosophical monstrosity."[6] Hugh Everett, as a fellow grad student and friend remembered, "always wanted to go away the winner," while "spending most time buried in a science fiction book."[7] Accordingly ambitious was his PhD project. Everett set out to do nothing less than to revise the mysterious quantum measurement process and apply quantum mechanics to the universe as a whole—and succeeded. He created what philosopher of science Max Jammer described later as "one of the most daring and most ambitious theories ever constructed in the history of science."[8] Ironically, Everett's theory became famous as the "many worlds interpretation" of quantum mechanics. In fact, it was an entirely monistic description of quantum reality.

Born on November 11, 1930, in Washington, DC, Hugh Everett III was an "army brat."[9] He would combine the left-brain capacities of his father, an engineer and US Army colonel, with the imaginativeness of his mother, who was flamboyant and liberal, a Romantic poet and science fiction writer. When Everett was five years old, the marriage of his parents broke apart. His mother left her husband, at first taking her son with her, until she couldn't cope with the many difficulties of single motherhood anymore. From the age of seven, Everett then lived with his father and, when his father fought in World War II in Europe, with his stepmom, Sarah. Growing up in times when patriotic citizens under the spell of Cold War paranoia would volunteer to watch the skies for Russian bombers or UFOs and built fallout shelters in their backyards in preparation for a nuclear attack, he was a chubby loner with a vivid imagination, a classic nerd who loved reading science fiction and playing with gadgets, with an insatiable appetite for practical jokes and logical paradox. "Life was a game to Hugh"; "he . . . was not used to living in the real world," Everett's later employee, physicist Gary Lucas, confirmed.[10]

When Everett was twelve years old, he wrote a letter to Einstein, hypothesizing about the paradox of what would happen when an irresistible force was acting on an immovable target. Einstein wrote back. "There is no such thing like an irresistible force and immovable body," he replied, "but there seems to be a very stubborn boy who has forced his way victoriously through strange difficulties created by himself for this purpose."[11] Six years later Everett graduated with honors from his military school and enrolled at the Catholic University of America, where he alienated his professors with what he claimed to be a "logical proof" about the nonexistence of God. At the same time, he was remembered as "by far the best student" his math professor ever had. Equipped with a "once in a lifetime recommendation," in the fall of 1953 Everett moved on to graduate school at Princeton University, whose famous neo-Gothic campus was within walking distance of the Institute for Advanced Study where Robert Oppenheimer, Einstein, and von Neumann worked.[12]

In Princeton Everett studied quantum mechanics using both von Neumann's and David Bohm's books. Bohm had gotten his PhD with Oppenheimer and had worked as an assistant professor at Princeton up until three years before Everett arrived. Bohm had been active in communist organizations and refused to testify against his colleagues, and as McCarthyism seized American politics, he was arrested and fired. In physics, he had been a fervent advocate of Louis de Broglie's approach to amend quantum mechanics with hidden variables, but with a sense of quantum mechanics' monistic implications. In Bohm's textbook, Everett could read sentences like "If the quantum theory is ... to provide a complete description of everything that can happen in the world ... it should also be able to describe the process of observation itself in terms of wave functions"; Bohm argued that while in classical physics "the world can be analyzed into distinct elements," quantum physics implies "the indivisible unity of all interacting systems."[13] Both insights would later become important cornerstones of Everett's own take on quantum theory.

Alongside his studies of quantum mechanics, Everett took an interest in game theory, the mathematical analysis pioneered by von Neumann and his Princeton colleague, economist Oskar Morgenstern, aimed at determining the best strategy to win a game, be it dealing with a conflict such as the Cold War or investing in the stock market. Attending lectures by Alfred Tucker, the PhD advisor of later Nobel laureate John Forbes Nash, subject of the 2001 Hollywood biopic *A Beautiful Mind*, Everett developed into a software whiz and game theorist. He would design computer simulations of nuclear wars, including a program for targeting cities with nuclear weapons, and a code-breaking algorithm for the National Security Agency. But most important of all, he revolutionized the foundations of physics and came up with the most truthful, yet most controversial, interpretation of what quantum mechanics implies for reality.

What's This Extra Jazz For?

In the fall of 1954, half a year after Everett had attended Einstein's seminar, Niels Bohr and his assistant, Aage Petersen, spent four months visiting the Institute for Advanced Study.[14] One night at the Graduate College, "after a slosh or two of sherry," Everett, Petersen, and the later quantum gravity pioneer Charles Misner started to ruminate about the meaning of quantum mechanics.[15] Everett amused himself by pointing out the outrageous implications of what Misner and Petersen said. "He loved to argue. I think it was his favorite sport," Misner remembered of his friend, and added, "When Niels Bohr visited Princeton, and his young assistant tried to explain Bohr's view on quantum mechanics, Hugh found it medieval: While mathematical formulated physics applied to everything when no one was looking, as soon as the results were to be unveiled God threw the dice."[16]

These jokes with friends soon turned into a thesis. After all, von Neumann, in his famous book titled *Mathematical Foundations of Quantum Mechanics*, had already summed up the theory as having "a peculiar dual nature . . . not satisfactorily explained."[17] In other words, the evolution of the wave function according to Erwin Schrödinger's equation didn't provide a unique outcome for a measurement, so it obviously didn't match with the experienced reality. In such a case, there are three possible options. One can argue that physics doesn't describe reality full stop; that was Bohr's approach. Or one can try to confine physics to what is experienced, as Bohm advocated. Or, finally, one can extend the notion of reality to what the equations of quantum mechanics produce.

Everett, as always fascinated by paradox and would-be realities, was hooked. As he remembered, he went to Wheeler and said, "Hey, how about this, this is the thing to do . . . [T]here is an obvious inconsistency in the theory."[18] "Let's just believe the basic equations—what's this extra jazz for?"[19] What Everett had in mind was an exclusively quantum mechanical approach to physics. As he explained to Petersen later, Everett believed that "we should no longer regard quantum mechanics as a mere appendage to classical physics tacked on to cover annoying discrepancies in the behavior of microscopic systems."[20] Instead, Everett was convinced that "the time ha[d] come to treat [quantum mechanics] in its own right as a fundamental theory without any dependence on classical physics, and to derive classical physics from it."[21] Everett proposed a quantum mechanical approach to everything that would start with a universal wave function for the universe: "So you do get a weird and funny picture," Everett rejoiced.[22]

With this approach, Everett struck a nerve with Wheeler, who considered himself a "radical conservative." Wheeler, as Misner described his mentor's philosophy, "was preaching this idea that you ought to just look at the equations and obey the fundamentals of physics while you follow their conclusions and give them a serious hearing."[23] Kip Thorne added that Wheeler had acquired this attitude from his own mentor, Niels Bohr: "Wheeler's often unconventional vision of nature was grounded in reality through the principle of radical conservatism, which he acquired from Niels Bohr: Be conservative by sticking to well-established physical principles, but probe them by exposing their most radical conclusions."[24]

Wheeler also had a practical reason for his interest in Everett's work. His focus on a universal wave function looked promising as a way to develop a theory of quantum gravity and cosmology. As Misner, who probably was the first person to use the phrase "quantum cosmology," remembers, "Everyone talking to Wheeler at that time was likely to be encouraged to think about quantum gravity."[25]

Within a year, Everett claimed he had solved the measurement problem. Everett's congenial solution assumed quantum mechanics to be universally valid and thus equally applicable to both elementary particles and to macroscopic objects such as the universe itself. To do so, Everett took quantum mechanics seriously as a theory about nature, not as a theory about knowledge, as the Copenhagen interpretation suggests. Nor did Everett add anything new to the theory. In contrast to Einstein, Everett didn't deem the theory to be "incomplete." As a consequence, he rejected both de Broglie's and Bohm's "hidden variables" and von Neumann's "collapse of the wave function." According to Everett, in a quantum measurement, all possible results are equally realized, albeit in different "relative states," in his own terms, or in parallel universes or "many worlds," in Bryce DeWitt's later rephrasing.

A particularly attractive feature of Everett's theory is how it deals with the Einstein-Podolsky-Rosen (EPR) paradox. Applying Everett's view to the example of mixed colors would imply that our original chemist and her container taken home split into many copies, one for each possible combination of paint—as does the other container left behind. Since the quantum potentialities do not collapse into a single outcome, no collapse needs to be transmitted with infinite speed. Only when the chemist walks back with her container—at a speed considerably slower than the speed of light—and compares it with the one in her lab do the matching parallel realities meld: the chemist who has seen dark blue paint in her bucket at home shares one reality with the light yellow bucket left behind, and the chemist who takes away a dark green bucket settles in a parallel reality with the white paint in her lab.

This allowed Everett not only to solve the EPR paradox but to make sense out of quantum mechanics by only using its quantum wave formalism—without resorting to the additional postulate of a collapse not included in the formalism. But Everett's idea comes with a price tag: it requires a multitude of parallel realities. A few years earlier, Schrödinger had pondered such implications, only to dismiss them at once: "The idea that . . . all really happen simultaneously seems lunatic . . . just impossible. . . . If the laws of nature took this form . . . we should find our surroundings rapidly turning into a quagmire, or sort of featureless jelly."[26] In contrast to Schrödinger, Everett wasn't intimidated by lunacy. In fact, he enjoyed it. And he had mathematics on his side. The quantum multiverse was born.

Many Worlds

According to quantum mechanics, during a measurement, as in any other interaction, the observer gets entangled with the observed object. In the case where the object is in a quantum superposition, such as a particle being "half here and half there," the observer gets split into two copies: one copy of the observer experiences the particle "here," and another copy observes the particle "there." "Why doesn't our observer see a smeared out needle [of the apparatus indicating the position of the particle]?" Everett asked, then clarified, "The answer is quite simple. When he looks at the needle (interacts) he himself becomes smeared out but at the same time correlated to the apparatus and hence to the system." As a consequence, "the observer himself has split into a number of observers, each of which sees a definite result of the measurement."[27] Thus, according to Everett, when a particle being in a superposition of two possible locations is observed, the particle doesn't collapse into one of the possible spots but rather the observer and its observation device split into two copies, one copy observing the particle in the first place and another copy finding the particle in the second place. Everett goes on to add, "Furthermore, should our observer call over his lab assistant to look at the needle, the assistant would also split, but be correlated in such a manner as always to agree with the first observer as to the position of the needle, so that no inconsistencies would ever arise."[28] Everett went as far as comparing the observer with an amoeba that reproduces via cell division, or "splitting," and concludes, "Our amoeba does not have a life line but a life tree."[29]

As a consequence, every single quantum process results in a multitude of observers, witnessing each possible outcome and thus living in their private individual realities, universes, or "Everett branches": "This universe is constantly splitting into a stupendous number of branches ... Moreover, every quantum transition taking place on every star, in every galaxy, in every remote corner of the universe is splitting our local world on earth into myriad copies of itself ... Here is schizophrenia with a vengeance," Bryce DeWitt dramatically wrote of this unsettling consequence of Everett's interpretation.[30] As Wheeler realized, "It is difficult to make clear how decisively the 'relative state' theory drops classical concepts. One's initial unhappiness at this step can be matched but few times in history," he writes, and goes on to compare Everett's theory with the revolutions initiated by Isaac Newton, James Clerk Maxwell, and Einstein: "No escape seems possible from this relative state formalism . . . [It] does demand a totally new view on the foundational character of physics."[31]

Yet, as much as Wheeler sympathized with Everett's radically conservative and entirely quantum mechanical approach, and as much as he valued Everett himself, he felt uncomfortable with the profusion of worlds and splitting observers. "Its infinitely many worlds make a heavy load of metaphysical baggage," was Wheeler's diagnosis.[32] In addition, there was a more personal reason for Wheeler to be torn about Everett's work: by all means Wheeler wanted to avoid getting drawn into a controversy with Bohr. According to DeWitt's wife, physicist Cécile DeWitt-Morette, "When [Wheeler] first saw the Everett paper, he was actually very uncomfortable because it was questioning Bohr."[33] As Misner explains, "Wheeler regarded Bohr as his most important mentor. He really adored Bohr."[34] "Bohr," Wheeler confessed, "taught me a new way of looking at the world."[35]

As soon as Everett finished his thesis, the problems began. "No one could fault his logic, even if they couldn't stomach his conclusions . . . The most common reaction to this dilemma was just to ignore Hugh's work," Misner determined.[36] In this situation, Wheeler was "frankly bashful about showing" Everett's draft to Bohr.[37] "In its present form, valuable and important as I consider it to be," parts of it may become "subject to mystical misinterpretations by too many unskilled readers," Wheeler believed.[38]

After Everett had submitted his thesis in January 1956, Wheeler wrote to Bohr to inquire about whether there might be an alternative to Everett's reasoning. In May Wheeler visited Copenhagen and discussed Everett's thesis draft with Bohr and Petersen, but obviously Bohr didn't like where Everett's work had taken him. Wheeler wrote back to Everett that "complete misunderstanding of what physics is about will result unless the words that go with the formalism are drastically revised" and that the thesis would need "a lot of writing and rewriting."[39]

At the same time he flattered Everett—"You (among the very few in the world) have the ability in thinking and in writing... you have it"—and urged him to discuss the matter out directly with Bohr: "Go and fight with the greatest fighter."[40]

But Bohr's and Petersen's criticism wasn't confined to wording. The Copenhagen physicists opposed the basic idea that quantum mechanics could be applied to macroscopic objects such as the measurement apparatus or an observer: "Silly to say apparatus has a wave function," Wheeler's notes of the discussion quote Petersen as saying.[41] When Everett read these notes, he just scribbled, "Nonsense!" on his own copy.[42] While Wheeler tried to appease the Copenhagen physicists, arguing now that "Everett's thesis [was] not meant to question the present approach to the measurement problem, but to accept and generalize it," Everett began losing interest:[43] "Hugh would, of course, have been happy if his quantum ideas were noticed and applauded, but when they were mostly ignored he was instead chagrined and perplexed," Misner remembered. "He could not understand why a perfectly logical idea had so little impact. But he had more important things to do than help the world properly understand quantum theory. He needed a job that would make lots of money and keep him out of the post–Korean War draft."[44] In June 1956, Everett left the university and took a top-secret job at the Pentagon. Meanwhile, struggling to find a way to reconcile Everett's work with Bohr's airy philosophy, Wheeler put Everett's thesis on hold and demanded a revision. After that, for more than a year and a half, nothing happened.

Things moved on only after Wheeler's group attended the Conference on the Role of Gravitation in Physics in January 1957, organized by Cécile DeWitt-Morette at the University of North Carolina in Chapel Hill. Everett didn't participate, but his theory was discussed. Besides Wheeler and the DeWitts, the attendants included Misner, who used this occasion to introduce the term "quantum cosmology" into the physics lexicon. Richard Feynman was there as well and articulated his incredulity about an "infinity of possible worlds" arising from Everett's universal wave function.[45] Right after the conference, Everett and Wheeler got together to rewrite the thesis that would later be published in the conference proceedings. Under Wheeler's surveillance, the thesis lost 80 percent of its content, and its title changed from "The Theory of the Universal Wave Function" to "Relative State Formulation of Quantum Mechanics." DeWitt, who edited the conference proceedings, was skeptical at first. While he described Everett's approach as "valuable" and "beautifully constructed," he objected, "I can testify to this from personal introspection . . . I simply do not branch."[46] Everett replied by referring to the adversaries of Copernicus's discovery that Earth was orbiting the sun: "I can't resist asking. Do you feel the motion of the earth?"[47] DeWitt relented and replied, "Touché!"[48] Over the following decade he developed into Everett's most fervent champion. Still, for the next thirteen years, Everett's work remained "one of the best kept secrets in this century," as philosopher of science Max Jammer put it.[49] DeWitt partly blamed Wheeler's revision of Everett's thesis: "The funny thing is, you have to read the . . . [rewritten thesis] very carefully . . . to see what's really there. Whereas in the Urwerk [the original version] it's quite well spelled out."[50] Not until 1973 was Everett's original, long thesis published by DeWitt and his PhD student Neill Graham in a book titled *The Many-Worlds Interpretation of Quantum Mechanics*.[51]

To make matters worse, the Copenhagen physicists' fierce opposition had not abated. In the spring of 1959, Everett had finally complied with Wheeler's admonition and traveled to Copenhagen to discuss his thesis directly with Bohr. But the "fight with the greatest fighter" fizzled. As Everett's biographer Peter Byrne describes, when Everett pitched his theory to Bohr and a few other physicists, including Bohr's collaborator, Belgian physicist Leon Rosenfeld, "there was simply a polite hearing and lot of mumbling," interrupted by Bohr relighting his pipe—"and that was it."[52] As Everett's wife, Nancy, remembered, "Bohr was in his 80s and not prone to serious discussion of any new (strange) upstart theory."[53] Rosenfeld's verdict was much harsher: "With regard to Everett, neither I nor even Niels Bohr could have any patience with him, when he visited us in

Copenhagen . . . in order to sell the hopelessly wrong ideas he had been encouraged, most unwisely, by Wheeler to develop. He was undescribably [sic] stupid and could not understand the simplest things in quantum mechanics."[54] Everett's own recollection was blunt: "That was a hell . . . doomed from the beginning."[55]

Frustrated, Everett retired to his hotel room, drank a lot of beer, and developed the powerful optimization method that he exploited successfully both in his warfare simulations to devise the US Cold War strategy and in his later consulting work. With very few exceptions, Everett would soon concentrate on his bread-and-butter job at the Pentagon, which would secure him "enough money to indulge himself with rich foods, fine wines, sexual escapades, and Caribbean cruises," as his biographer Peter Byrne wrote.[56]

While Wheeler still believed that Everett's talent was wasted in the military industry and carried on with convincing several universities to offer him academic jobs, Everett didn't show any interest. Everett rarely wanted to talk about quantum mechanics anymore. When his later friend and business partner, physicist Donald Reisler, first applied for a job to work with Everett in 1970, Reisler remembered Everett asking him shyly whether he ever had heard of the relative states theory. "Oh my God, you are that Everett, the crazy one," Reisler thought.[57] They never spoke about the topic again, and when three years later they founded a company together, they agreed to lock up their dissertations in a file drawer and not to talk about quantum mechanics for the next ten years until they "presumably . . . could afford the luxury of such a diversion."[58]

At the same time, Rosenfeld had started a crusade against Everett and everyone else who dared to question the Copenhagen orthodoxy. As a Marxist, he had recognized the parallels between Bohr's complementarity and "dialectics," the idea that the pros and cons of disagreement can be resolved through a synthesis of arguments. While dialectics can be traced back all the way to Heraclitus and Plato, it was later adopted by Karl Marx and Friedrich Engels as a centerpiece of the political philosophy of Marxism. As a consequence, complementarity to Rosenfeld became a matter of ideology. According to historian of science Anja Skaar Jacobsen, Rosenfeld took "up the fight against all disbelievers of complementarity whether . . . Marxist physicists or just supporters of the causal program with no Marxist agenda. It was a fight in which he used all possible means, including polemical papers, book reviews and personal connections."[59] Rosenfeld served as a consultant or referee for several important publishers and the prestigious journal *Nature*, and in this capacity he took care to make sure that ideas questioning the orthodox Copenhagen philosophy were suppressed. Especially Everett's work, Rosenfeld claimed, "suffer[ed] from fundamental misunderstanding," was "perfectly trivial, but also terribly treacherous," and led Everett to a conclusion that was "an illusion."[60]

Finally, as Misner observed, in any case "quantum physicists had their hands full around 1957 with exciting research that found Bohr's viewpoint adequate." As Wojciech Zurek, who got his PhD with Wheeler twenty-two years after Everett and became one of the pioneers of decoherence theory, describes this pragmatic attitude, "Instead of trying to understand the Universe (including 'the classical') in quantum terms, one 'quantized' this and that, always starting from the classical base."[61] Misner details how "new elementary particles were being discovered and their relations systematized . . . [N]uclear structure was beginning to make sense . . . as was the source of energy in the Sun; superconductivity had just been explained, and condensed matter theory was flowering supported by the success of the transistor. None of this would benefit by using Hugh's view of the quantum instead of Bohr's."[62] Everett agreed with this judgment: "Unfortunately, as it turned out, the theory which I constructed resolved all the paradoxes and at the same time showed the complete equivalence with respect to any possible experimental test of my theory and that of conventional quantum mechanics. The net result of my theory therefore is simply to give a complete and self-consistent picture (without any particular 'magic' associated with the measurement)."[63] Yet, Everett maintained, "if one will only swallow the world picture implied by the theory, one has, I believe, the simplest, most complete framework for the interpretation of quantum mechanics today."[64]

From a modern perspective though, there indeed are important consequences of the Everett interpretation, particularly if one wants to make sense of a quantum mechanical description of the universe or of quantum computing, one of the most vibrant research fields in physics today. David Deutsch, a pioneer of the field of quantum computing, was a PhD student of Dennis Sciama (as Stephen Hawking had been about fifteen years earlier) when he met with Everett in a beer garden in Austin in 1977. Everett had just delivered a lecture, capturing Deutsch's imagination.[65] Deutsch explains the importance of Everett's insights by noting that quantum computers benefit from the fact that they perform calculations simultaneously in different parallel worlds. "I won't say 'Everett's interpretation,' but 'Everett's theory'—which is quantum theory," Deutsch insists, as "no single-universe theory can explain even the Einstein-Podolsky-Rosen experiment, let alone, say,

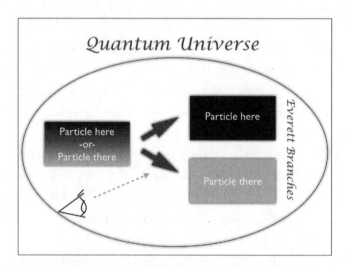

In the Everett interpretation, during a measurement or observation the universe splits into various branches. From a global perspective, though, all branches are still part of the quantum universe.

quantum computation."[66] To emphasize the dramatic dimensions of this insight, he writes, "When a quantum factorization engine is factorizing a 250-digit number, the number of interfering universes will be of the order of 10^{500}—that is, ten to the power of 500 . . . All those computations are performed in parallel, in different universes, and share their results through interference."[67] While Everett's theory doesn't provide any experimental signatures different from the Copenhagen interpretation, it is essential to make sense of modern physics, from cosmology to its most advanced technical applications. Yet, despite such virtues, Everett's interpretation remained an outsider's view.

From Many Worlds to One

There is a core contradiction to be resolved as we explore what Everett's theory has to say about a monistic view of the universe. It may seem that by trying to take quantum mechanics seriously as a theory describing the reality of nature, Everett sacrificed the uniqueness of the universe and replaced it with a multiverse of many worlds. But this conclusion results from a superficial consideration. What is typically overlooked in this picture is that Everett's multiverse is not fundamental but rather apparent or "emergent," as philosopher David Wallace at the University of Southern California insists.[68] From a fundamental perspective, rather than splitting the universe apart, Everett's formalism allows application of quantum mechanics to the entire universe and thereby enables entanglement to merge the universe into an all-encompassing "One." Everett put this point straight, when asked about it in 1977 by the young French physicist Jean-Marc Lévy-Leblond: "The question is one of terminology: to my opinion there is but a single (quantum) world, with its universal wave function. There are not 'many worlds,' no 'branching,' etc., except as an artifact due to insisting once more on a classical picture of the world."[69]

In his letter, Lévy-Leblond included a paper he had presented one year earlier at a conference held in Strasbourg that further explained his point. As Lévy-Leblond points out, Everett's interpretation "is said to describe 'many universes,' one for each of these branches." The difference between Copenhagen and Everett is then typically illustrated as follows: "Where the Copenhagen interpretation would arbitrarily choose 'one world' by cutting off all 'branches' . . . except one (presumably the one we think we sit upon), one should accept the simultaneous existence of the 'many worlds' corresponding to all possible outcomes of the measurement." While the defining feature of Everett's interpretation is that it describes the universe exclusively in quantum mechanical terms, yet, as Lévy-Leblond emphasizes, the eponymous "many worlds" describe classical realities: "the 'many worlds' idea again is a left-over of classical conceptions," in obvious contradiction of Everett's original intent, as Lévy-Leblond finds. "To me, the deep meaning of Everett's ideas is not the coexistence of many worlds, but on the contrary, the existence of a single quantum one," Lévy-Leblond writes in summarizing his criticism of how Everett's interpretation is usually characterized.[70]

In his reply, Everett agrees: "Your pre-print . . . is one of the more meaningful papers I have seen on the subject, and therefore deserving of a reply." He clarifies that "the 'Many-Worlds Interpretation' . . . of course, was not my title as I was pleased to have the paper published in any form anyone chose to do it in!" and confirms Lévy-Leblond's conclusions: "But your observations are entirely accurate (as far as I have read)."[71]

Everett reinforced what he had already written in an early paper describing his dissertation project to Wheeler—"The physical 'reality' is assumed to be the wave function of the whole universe itself"[72]—and what he had explained in his long thesis: "It is meaningless to ask about the absolute state of a subsystem—one can only ask the state relative to a given state of the remainder of the system."[73] About this point, Everett completely agreed with David Bohm, who had emphasized already in his 1951 textbook that "at the quantum level of accuracy the entire universe must be regarded as forming a single, indivisible unit with every object linked to its surroundings."[74] In fact Everett's original "The Theory of the Universal Wave Function" became the "Relative State

Formulation of Quantum Mechanics" under the influence of Wheeler, only to be later renamed the "many universes" or "many worlds" interpretation by DeWitt. Step-by-step the focus on a monistic, entirely quantum mechanical description of the universe was shifted toward an obsession with parallel existing classical realities. The late Oxford philosopher Michael Lockwood goes as far to call DeWitt's label "a distinct misnomer for the view."[75]

In contrast, the fundamental single quantum universe not only alleviates the purported problem of Everett's interpretation featuring a "heavy load of metaphysical baggage" (as Wheeler complained) but totally invalidates this criticism as such a fundamental reality isn't only a single universe but a single unique entity comprising matter, space, and time as well as all potentially possible events and situations. Not only is there just a single world, but this single world is all there is! Little known as it is, this consequence of his theory may turn out to be Everett's paramount legacy. As Wojciech Zurek affirmed, "It was Everett who gave us the permission to think about the universe as wholly quantum mechanical."[76]

It took decades, though, until Everett's idea was taken seriously as an interpretation of quantum mechanics. One reason was the pragmatic "Shut up and calculate" attitude of the physicists at the time. Another was the fierce resistance of Bohr and his collaborators. Beyond that, however, there was the problem that it wasn't quite clear how Everett's multiverse was related to our experienced, classical reality.

Down-to-Earth: The Brunswickan

If, as Everett emphasized, his theory treated the measurement "as a natural process within the theory of pure wave mechanics," and if, as he affirms, during a measurement, "due to interactions, strong correlations will be built up" with the environment, and if "it is this phenomenon which accounts for the classical appearance of the macroscopic world, the existence of definite solid objects, etc."—how exactly does this process actually happen?[77]

It would take more than ten years and the work of a physicist quite different from Everett to answer these questions and to put Everett's theory on a firm footing; a physicist who grew up in a war-ravaged country while Everett developed computer programs to simulate nuclear wars; a down-to-earth, sober soul as opposed to the sex-obsessed, chain-smoking, alcoholic Everett. This physicist was a bulldog about the foundations of physics, whereas Everett, frustrated about the opposition and neglect his theory faced, "in effect, had washed [his] hands of the whole affair in 1956" and locked up his thesis to avoid ever being tempted to talk about it.[78] After having been frankly informed by his advisor, a famous Nobel Prize winner, "that any further activities on this subject would end [his] academic career!" this physicist just wouldn't give in. Instead, he thought, now that his "career was destroyed" anyway, he might just as well enjoy his jester's license: "Now I can just do what I like and I don't have to try any more to get any position, or something like that."[79]

Heinz-Dieter Zeh was born in Braunschweig, aka Brunswick, Germany, in 1932. Among Germans, people from Brunswick are generally regarded as grounded, reserved, humble, and outspoken, a characterization that seems to describe Zeh himself pretty well. Zeh was less than a year old when the Nazis seized control of the country; he was seven years old when the sanguinary dictatorship instigated World War II; and on his thirteenth birthday, the regime collapsed, and the war ended in Europe with Germany's unconditional surrender. In the months before, about 90 percent of Brunswick's inner city, dating back to the medieval ruler Henry the Lion, had been reduced to ashes by Allied air raids, while the Nazi Party sent children only three years senior to Zeh himself to the front lines of a lost war. If the Germans couldn't win, Hitler reportedly argued, then they deserved to perish. As if this weren't enough, after Germany was split into two states, one under Soviet influence, as a consequence of the defeat, Brunswick would suffer economically for almost half a century due to its geographical proximity to the Iron Curtain separating the Western and Eastern zones of influence in Europe. Did Zeh's background and his childhood mold him in a way that made him particularly unwilling to accept preconceived notions and follow party lines? This is hard to say. As his widow, Sigrid Zeh, emphasized, he "was very reluctant to tell about his life."[80]

Fact is, when Zeh started his career in physics, the party line for quantum mechanics was the Copenhagen interpretation, and Zeh was everything but willing to comply: "I expect the Copenhagen interpretation will sometime be called the greatest sophism in the history of science," he wrote to Wheeler in 1980.[81] For Zeh, whose main motivation was "the search for a unified and conceptually consistent description of nature in modern physics," the Copenhagen interpretation was simply a surrender of consistency.[82]

Zeh had started studying physics in Braunschweig but soon moved to the University of Heidelberg, where he could work with Nobel laureate Hans Jensen. Jensen was famous for the shell model of the atomic nucleus that was developed independently also by the German-born American physicist Maria Goeppert Mayer (for which they shared the 1963 Nobel Prize). According to the shell model, the atomic nucleus resembles a miniature atom, with the difference that its orbits are occupied by protons and neutrons instead of electrons. It was in the analysis of such nuclear models that Zeh made a groundbreaking discovery of his own.

After the completion of his PhD, Zeh had spent some time as a postdoc at the University of California (UC), Berkeley, the California Institute of Technology, and UC San Diego, before he returned to Heidelberg in the mid-1960s to prepare his "habilitation" thesis, which would qualify him to apply for professor positions in Germany. For the young physicist, the 1960s must have been a time of change and promise. He left his gray, depressing home country, still struggling to overcome the devastations of war, to live in sun-kissed California, then returned to a Germany where students were protesting in favor of free love and political participation and against the Vietnam War and former Nazi bureaucrats still occupying high-ranking positions in the German administration. "Zeh wasn't as uptight as his conservative peers among Heidelberg's physics professors," Zeh's early student Bernd Falke remembers.[83] Instead, "Zeh was open-minded and critical about the conservative politics. He was cruising around in a white Porsche equipped with a hitch to tow his sailboat and came to our parties where the students discussed and smoked marihuana."[84] At some point, Zeh traded his Porsche for a Mercedes. "He must have met his later wife and wanted to appear more respectable," Falke suspected.[85] And while down in the Neckar valley the conflict between revolting students and the authorities escalated, while the Institute for Political Science got occupied and the university's president threatened to shut down the university, and while the violence between rioting protesters and the police intensified, Zeh was working in his office, at the Philosophenweg, halfway up the mountains bordering the northern riverbank, propelling his own revolution.

In these days, Zeh worked with common approximation methods for nuclear physics that described the atomic nucleus in terms of its constituents. Such considerations led Zeh to think about the relations of quantum subsystems to the whole: "This analogy . . . led me to entertain the wild speculation of a gigantic atomic nucleus, big enough to contain complex subsystems such as measurement devices and even conscious observers," Zeh later remembered of his first foray into quantum cosmology.[86]

It was this conception of the universe as a gargantuan atomic nucleus that inspired Zeh to ask a pivotal question: What would such a nucleus look like from the perspective of the protons and neutrons, the particles comprising it? How, in other words, would the quantum universe be experienced by the observer within? Eventually, these questions led Zeh to discover the solution to the measurement problem and the phenomenon that would later be called decoherence.

Decoherence acts as the agent protecting our daily-life experience from too much quantum weirdness, resolving the conundrum of how classical experience emerges in a quantum measurement. Whenever a quantum system is measured or coupled with its environment, entanglement causes the quantum system, the observer, and the rest of the universe to become interwoven with each other. Consequently, from the perspective of the local observer, who can't oversee the entire universe, information is dispersed into her unknown environment. This information that appears to be lost from the perspective of the observer is the glue between quantum realities. Without it, quantum mechanical superpositions—such as Schrödinger's infamous undead cat—break up into parallel realities in which quasi-classical objects such as particles with a definite location emerge. As a consequence, decoherence acts as if it would open a zipper between quantum physics' parallel realities. From the observer's perspective, the universe and she herself seem to "split" into separate Everett branches. The observer observes a live cat or a dead cat but nothing in between. The world looks classical to her, while from a global perspective it is still quantum mechanical. In fact, in this view the entire universe is a quantum object. The situation is just like that described in a famous German lullaby—"Behold the moon—and wonder, why half of her stands yonder, yet she is round and fair. We are the ones who're fooling, 'cause we are ridiculing, as our minds are unaware"—with the full moon representing the quantum universe and its visible half-moon image playing the role of our experienced, classical Everett branch.

When Zeh had finished the preliminary draft of his paper, he left his office and climbed down the steep walk to the institute's library to do something "completely different."[87] In the beautiful mansion at the Philosophenweg, overlooking the city of Heidelberg and the Neckar valley, Zeh stumbled on something rather similar instead: an article by Bryce DeWitt on quantum gravity that made use of Everett's formalism. As Zeh realized immediately, what he had discovered was the missing piece underlying Everett's "many worlds interpretation": the agent of the birth of what we call reality.

How Matter Is Born

Among the most striking consequences of decoherence is that matter, understood as solid stuff made out of particles, may turn out to be an illusion. Zeh titled a 1993 paper "There Are No Quantum Jumps, nor Are There Particles!"[88] In his review "The Strange (Hi)story of Particles and Waves," which he constantly updated, with twenty-three versions between 2013 and his death in 2018, he further affirmed, "The particle concept was recognized as a delusion."[89]

The basic idea behind this notion is that matter—or, more specifically, particles—is not fundamental; it is "emergent," as philosophers describe the nature of concepts that are useful for practical purposes but do not exist on a closer look, such as, for example, "temperature," which boils down to the average energy of atoms or molecules from a microscopic perspective. As Zeh explains for the example of photons, the quanta of light and electromagnetic radiation, "The spontaneous occurrence of photons as apparent particles (in the form of clicking counters, for example) is then merely a consequence of the fast decoherence caused by the macroscopic detector."[90] Zeh's student Erich Joos concretized later, "'Particles' appear localised in space not because there are particles, but because the environment continually measures position. The concept of a particle seems to be derivable from the quantum . . . state."[91] According to Zeh and Joos, what looks like matter emerges via decoherence from the quantum mechanical wave.

Zeh's discovery is reminiscent of Schrödinger's critique of Werner Heisenberg's particle picture. In two essays written in 1952 for the *British Journal for the Philosophy of Science*, Schrödinger had argued that the eponymous "energy parcel view" of quantum mechanics "is an illusion" and that "one is allowed to regard microscopic interaction as a continuous phenomenon without losing . . . any . . . understanding of phenomena that the parcel-theory affords."[92] Like Zeh, Schrödinger believed that quanta are not fundamental but a consequence of our sloppy description of nature: "When we hear the same words again and again pronounced with authority, we are apt to forget that they were originally meant as an abbreviation; we are induced to believe that they describe a reality."[93] But where Schrödinger shied away from the "quagmire" or "jelly" entailed by an entirely quantum mechanical foundation of physics, Zeh emphasized that for him "the most important fruit of decoherence (that is, of a universal entanglement) is the fact that no classical concepts are required any more on a fundamental level."[94] Zeh further demonstrated why Everett's universal wave function could look as though there actually were particles. In other words, Zeh explained how, if all is One, One still can appear as many things.

To fully appreciate this argument, we get back to Zeh's first discovery of decoherence when he worked on a description of nonspherical, "deformed" atomic nuclei and applied the standard approximation methods of nuclear physics. While the formalism worked perfectly fine, Zeh was deeply puzzled by the logic justifying these approaches. Here, typically, a composed total system (in Zeh's case, the atomic nucleus) was described by a superposition of individual components (in Zeh's case, the protons and neutrons making up that nucleus) that had features the total system didn't. For example, the nucleons could rotate in a certain way, while the entire nucleus would remain steadfast. This went as far as approximating a time-independent nucleus in terms of time-dependent components: "According to which logic can the solution of a time-independent equation approximately depend on time?" Zeh asked himself.[95] On the other hand, strictly speaking, these individual components didn't exist in well-defined states. As subsystems of an entangled total system, the components were entirely merged into the whole. Zeh realized that by looking at a subsystem, one could experience properties that were not existent in the complete description of the total system. If nucleons inside the nucleus "feel a definite [property] in spite of" this property being completely absent in the nucleus considered in its entirety, "would an internal observer then not similarly have to become 'aware of' a certain measurement result?" Zeh wondered.[96]

To make a long story short, decoherence allows the observer to experience things that are not really existent, as a consequence of his limited information about the whole. While we usually see less when we overlook something, decoherence perplexingly allows us to see more than what really exists. To get an intuitive grasp of this paradoxical behavior, we can again resort to the metaphor of the projector reality. In fact, at the heart of the projector there isn't the film roll but a featureless light source. In an early predecessor of the film projector, the magic lantern, pictures painted on glass slides placed between the light bulb and the projection screen absorbed some of the light emitted from the source. An even earlier and more basic analogy for the principle of decoherence is shadow play. In any of these cases, light from the light source is absorbed by the puppets or the pictures on the slide or the film roll to create the images displayed on the screen. The characters, objects, and stories we experience by watching the screen are a consequence of us *not* seeing all the light emitted from the lightbulb. Another example is the action of a colored optical lens. While it seems that such a lens adds color to the colorless sunlight, in truth the lens works by absorbing all other component colors present in the white, colorless state. Just like decoherence, a colored lens, a shadow puppet, a painted slide, or a film roll seemingly creates information by actually filtering out information. In all these cases, it is our ignorance that constitutes our experience.

This implies, however, that the components of a quantum system don't really exist: "The only object that can truly exist is the quantum state of the entire universe," Zeh concluded as early as 1967.[97] For Zeh it is the projector reality that is fundamental, and it constitutes an all-encompassing unity, an entangled "quantum universe."[98] Everything else, including matter or particles, is an illusion. This nonfundamental nature of matter is most dramatically uncovered in an effect found by John Wheeler's student William Unruh and other researchers in the early 1970s. According to the "Unruh effect," which is of crucial importance in making sense of black holes, an accelerated observer finds particles in empty space, the vacuum, while an observer at rest or constant speed sees nothing. The very existence of particles depends on the motion, or—more generally—the perspective of the observer. This demonstrates clearly that the existence of matter is a derived concept that may appear quite dissimilar for different observers.

Quanta Getting Big

Physics is an experimental science. And by the end of the day, it is experiments rather than lofty ideas that separate fact from fiction. If Everett and Zeh are right, it should be enough to just isolate any object from its environment to avoid decoherence and demonstrate that it is quantum rather than classical deep within. If this works, we can create quantum objects that are larger than particles or atoms, being as big as the objects in our everyday lives. According to philosopher Michael Lockwood, "The whole point of Schrödinger's cat example . . . is to show that, given a suitable coupling, any microscopic superposition can be made to generate a corresponding macroscopic one." As Lockwood explains, "There is nothing in the character of our ordinary experience that constitutes a shred of evidence that quantum mechanics does indeed break down at the macroscopic level. What is inconsistent with the universal applicability of quantum mechanics is not our ordinary experience as such, but the common-sense way of interpreting it."[99] Still, the lack of "any shred of evidence" that quantum mechanics breaks down for large objects is one thing. Much more convincing evidence for the universal validity of quantum mechanics would be to actually produce macroscopic quantum objects.

As it happens, starting from the 1990s, experimental physicists have indeed demonstrated that increasingly large quantum phenomena can exist. "It's not easy," warns American science writer Stephen Ornes. "Quantum effects are fleeting, delicate, and fragile, drowned out by even the slightest vibration or thermodynamic fluctuations. To observe them at all requires experimental setups that isolate the system from the heat and noise of the outer world."[100] Yet, by using electric currents circulating in a superconducting wire, it can be done. "Physicists can use magnetic fields to induce current to flow in both directions around the ring at the same time. That doesn't mean half go one way and half go the other; all the electrons . . . simultaneously stream clockwise and counterclockwise," writes Ornes.[101]

Other large quantum systems that have been realized involve large molecules, membranes, or entanglement between faraway places. For example, in 1999, a group of researchers around the Viennese physicists Anton Zeilinger and Markus Arndt demonstrated successfully the interference of quantum waves for so-called "Bucky Balls" or fullerenes, soccer-ball-shaped macro-molecules made out of sixty to seventy individual carbon atoms. In 2012, Zeilinger's group went on and transmitted quantum properties via entanglement, a process known as "quantum teleportation," over 143 kilometers between the two Canary Islands of La Palma and Tenerife. Another approach, pursued by Simon Gröblacher at Delft University in the Netherlands in 2016, achieved entanglement, using membranes one millimeter in diameter. Gröblacher even dreams of putting a living organism, such as a "tardigrade" or "water bear," into a quantum superposition. And in fact, in December 2021 an international group of scientists including physicists from Oxford and Singapore reported that they had succeeded in realizing entanglement between a live tardigrade and a superconducting solid-state device, implying that each subsystem exists in a superposition.[102]

Moreover, the study of macroscopic quantum phenomena isn't confined to observing superpositions; it includes observing "the gradual action of decoherence and thus the step-by-step transition between the quantum regime and the classical domain," as Maximilian Schlosshauer writes in his book *Decoherence and the Quantum-to-Classical Transition*.[103] To give an example, Schlosshauer describes how "interference patterns produced by large fullerene molecules sent through diffraction gratings were observed to decay gradually as the density of surrounding gas molecules, and thus the rate of scattering events between the fullerenes and the environmental particles, was increased."[104] As Schlosshauer emphasizes, "We are now in a position to directly measure how the continuous interaction with the environment gradually degrades our ability to observe quantum phenomena."[105] "Quantum theory is here to stay," determines Wojciech Zurek, adding, "It is also increasingly clear that its weirdest predictions—superpositions and entanglement—are experimental facts, in principle relevant also for macroscopic objects."[106]

Nowhere Land

But why did it take so long to appreciate the universal nature of quantum mechanics and the fundamental role of entanglement and decoherence? Why—in view of the long history of quantum mechanics—was it that such dramatic consequences for our notion of reality weren't realized earlier? "Zeh felt that the key to the problem lay with accepting the fundamental role that entanglement plays in quantum theory . . . not merely as a statistical correlation . . . but as a feature of the underlying 'reality,'" writes philosopher of science Kristian Camilleri.[107] Elsewhere, Zeh emphasized that he was "indeed convinced that the importance of decoherence was overlooked for the first 60 years of quantum theory precisely because entanglement was misunderstood as no more than a statistical correlation between local objects."[108] In fact, a straightforward consequence of entanglement and Zeh's insight that the apparent granular structure of matter, the organization of bodies in terms of atoms and particles, is an illusion is the "nonlocality" of quantum states. In general, quantum objects quite literally have no place in the universe—a concept that was rather alien at the time Zeh was pondering it though. "In my early works in English language the word 'entanglement' doesn't arise at all, simply because the term was so uncommon that I didn't even know the English translation of Schrödinger's German term 'Verschränkung,'" Zeh recalled later.[109] "Even when Schrödinger later called entanglement the greatest mystery of quantum theory, he used the insufficient phrase 'probability relations in separated systems.'" Zeh pointed out how the true significance of entanglement had been missed. "The importance of entanglement for the . . . binding energy of the Helium atom was well known by then," Zeh emphasizes.[110] Nevertheless, "Einstein, Podolski and Rosen . . . von Neumann . . . none of these great physicists was ready to dismiss the condition that reality must be local (that is, defined in space and time)."[111]

Around the same time that Zeh discovered decoherence in Heidelberg, in Geneva John Bell and Bernard d'Espagnat were racking their brains about related problems. In the 1960s, both d'Espagnat and Bell worked as theoretical particle physicists at the European Organization for Nuclear Research (CERN). "At that time, I had problems with quantum mechanics and he had problems with quantum mechanics. But he did not know I had and I did not know he had. I was once told by a friend that, for some reason or other, he was sort of suspected of having some."[112] A book finally betrayed Bell: "I had a confirmation of this when I spotted a heretical book on John's bookshelf," d'Espagnat recalled.[113] He and Bell began talking about the foundations of quantum mechanics. One of the starting points for the discussions was the bewilderment about entanglement: "I rediscovered for myself what, in fact, Schrödinger had ... pointed out a long time before ... something that I had not read about in any book, namely nonseparability"—the impossibility of splitting an entangled quantum state into components without losing information. "When I spoke of this with John he agreed of course."[114]

Just like Bohm, Everett, and Zeh before them, d'Espagnat and Bell soon encountered the same toxic blend of hostility and dogmatic pragmatism from their peers. As d'Espagnat found, "The trouble is that a great number of the physicists who came after Bohr clung to the formalism ... developed in Copenhagen and resisted any attempt at changing anything, but at the same time they adopted quite a different motivation principle, namely the one of scientific realism. In so doing they gave up consistency."[115] Bell agreed: "The typical physicist feels that these questions have long been answered and that he will fully understand just how if ever he can spare twenty minutes to think about it."[116] As physicist Andrew Whitaker described, "It was considered totally inappropriate even to think with any novelty about the fundamental nature of quantum theory and its rather surprising properties. That, it was practically universally believed, had been sorted out once and for all by Niels Bohr thirty years before."[117] John Clauser, who later performed the first experimental test of Bell's work, confirmed that "a very powerful . . . stigma began to develop within the physics community towards anyone who sacrilegiously was critical of quantum theory's fundamentals." This prejudice had consequences, and not only for a researcher's reputation: "The net impact of this stigma was that any physicist who . . . seriously questioned these foundations . . . was immediately branded as a 'quack,'" and "quacks naturally found it difficult to find decent jobs within the profession," Clauser added.[118] Whitaker agreed that while "a few brave souls did put their heads above the parapet . . . they were severely criticized by the powerful group of physicists centred around Bohr, Heisenberg, and Wolfgang Pauli. It was made fairly clear that, to be frank, there was no place in physics—no jobs in physics!—for anybody who dared to question the Copenhagen position."[119]

Thus Bell pursued his work on the meaning of quantum mechanics more or less secretly in his free time: "I am a Quantum Engineer, but on Sundays I have principles," Bell said of this working philosophy.[120] The most remarkable outcome of Bell's leisure activity was a reformulation of the EPR paradox in terms of an inequality: if quantum mechanics were correct as it was, if it was indeed nonlocal, then Bell's inequality should be violated. Even better, Bell's acid test for nonlocality wasn't confined to hidden variable versions of quantum mechanics. Bernard d'Espagnat recalls, "The clue is even deeper, since the experimental violation of the Bell inequalities proves nonlocality quite independently of whether quantum mechanics is right or wrong. This of course is most important."[121] As d'Espagnat emphasizes, "John brought a basic question, that of locality, down from the cloudy heavens of philosophy to the more solid realm of scientific research."[122] "Surprisingly, while quantum mechanics was already well established at the time of the publication of Bell's theorem, no experiments existed which definitely allowed one to rule out a local realistic interpretation," Bell's friend Reinhold Bertlmann and quantum information pioneer Anton Zeilinger later remembered. But this would change soon. Bell's ideas struck a nerve with a group of young iconoclasts in the United States who derived a generalization of Bell's theorem that was suitable for experimental testing, now known as the "Clauser-Horne-Shimony-Holt (CHSH) inequality." This reformulation allowed particle physicists to actually put entanglement and the ensuing nonlocality of quantum mechanics to a test to determine experimentally whether quantum mechanics implies that the whole is more than its parts, that the total spin of a pair of particles isn't located in the component spins of the individual particles. "The conclusions are philosophically startling; either one must totally abandon the realistic philosophy of most working scientists, or dramatically revise our concept of space-time," John Clauser, the C in "CHSH," recalled.[123] This holistic, "all is One" feature of reality finally proved that properties aren't necessarily located in specific places. They can be "nonlocal," such as the properties of Zeh's nuclei that can't be found in the component nucleons.

But hadn't it been one of Everett's major triumphs that he had proved the EPR paradox to be "fictitious"; that he could reconcile quantum mechanics with relativity by pointing out that measurement interactions could be local; that no superluminal information transfer was required to explain what happened during a measurement of an entangled pair of particles? This is indeed the case; yet the fact that Everett's interpretation allowed for local interactions and measurements didn't imply that under any circumstances properties of quantum states could be localized. It was this second type of nonlocality that Bell's inequalities proved: that a property of an entangled quantum state couldn't be reduced to the properties of its constituents simply because these constituents didn't exist as long as the total state was considered as a whole. This finding is in perfect agreement with Zeh's finding that quantum mechanics doesn't support the existence of particles (i.e., localized lumps of matter).

Bell's work established that the fundamental quantum reality is nonlocal, that it exists in a "nowhere land" beyond space and time. But it also had an effect on the general intellectual climate, on how physicists interested in the foundations of quantum physics were appreciated among their peers. As the historian of science Olival Freire Junior describes, "Three events about 1970 can evidence that change of mood."[124] The first one was a review of Everett's interpretation aimed at a general audience, published by Bryce DeWitt in *Physics Today*. Then there was the launch of the scientific journal *Foundations of Physics*, which published work on the interpretation of quantum mechanics, including Zeh's original paper on decoherence. And finally, in the summer of 1970, there was the International School of Physics "Enrico Fermi" in Varenna, a picturesque Italian village situated on a mountain ledge reaching into Lake Como. According to Freire Junior, "Varenna was the Woodstock of quantum dissidents."[125]

Meanwhile, Zeh in Heidelberg was still struggling to get attention for his work on decoherence. "It was absolutely impossible at that time to discuss these ideas with colleagues, or even to publish them," Zeh recalled later.[126] When he had written his paper, his mentor Jensen told Zeh that he did not understand the work. Of all people in the world, Jensen sought advice from Rosenfeld, who had already been sabotaging Bohm's and Everett's work before. Rosenfeld's answer was as merciless as it was wrong: "I have all the reasons in the world to assume that such a concentrate of wildest nonsense is not being distributed around the world with your blessing, and I think to be of service to you by directing your attention to this misfortune."[127] As Zeh remembered, "Jensen never showed me [Rosenfeld's letter] where he must have been very cynical about what I had said, and I remember that Jensen told that to some other colleagues, then when I noticed they were talking about them, they were chuckling. But he never told me precisely what was in this letter . . . Then Jensen told me that I should not continue this work, and so then our relationship deteriorated."[128] The only well-known physicist who reacted positively to Zeh's breakthrough was Eugene Wigner, a friend and colleague of John Wheeler and John von Neumann in Princeton who had shared the 1963 Nobel Prize with Jensen and Goeppert Mayer. "He helped me to get it published, and he also arranged for an invitation to a conference on the foundations of quantum theory to be held at Varenna in 1970,"— the "quantum Woodstock"—"organized by Bernard d'Espagnat," Zeh recalled.[129] Indeed, speakers in Varenna included Bell, Bohm, Wigner, de Broglie, DeWitt, Abner Shimony (the S in CHSH), and Alan Aspect, who later closed loopholes in the results proving the violation of Bell's inequalities. It also was in Varenna that DeWitt announced publicly his recent conversion to the Everett interpretation.

Yet, when Zeh arrived at the school, he was disappointed:

> I found the participants—John Bell included—in hot debates about the first experimental results regarding the Bell inequalities, which had been published a few years before. I had never heard of them, but I could not quite share the general excitement, since I was already entirely convinced that entanglement (and hence non-locality) was a well founded property of quantum states, which in my opinion described reality rather than probability correlations. So I expected that everybody would now soon agree with my conclusions. Obviously I was far too optimistic.[130]

Now Zeh had to listen to the other participants "searching for loopholes in the experiments which confirm the violation of these inequalities . . . even though all experimental results so far were precisely predicted by quantum theory."[131] Zeh couldn't "see anything but prejudice in such an assumption about reality."[132]

Still, Zeh didn't give up. As hard as it is to imagine Zeh publishing in obscure journals such as the *Epistemological Letters* or attending the 1983 workshop at the Californian hippie resort in Esalen, where the participants would sit naked in hot pools while discussing parapsychology and the physics of superluminal travel, he felt compelled to do so to advocate a rational foundation of reality. It took until 1991, when Wojciech Zurek introduced the concept in a classic *Physics Today* article, that decoherence eventually won recognition among the wider community of quantum physicists.[133]

Of Frogs and Birds

"Decoherence is—in my humble opinion—one of the most important discoveries of the past century," Massachusetts Institute of Technology

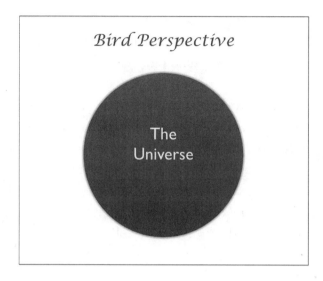

Bird Perspective

As experienced from the bird perspective, the universe is One.

cosmologist Max Tegmark wrote to me in an e-mail in 2017.[134] Beyond any doubt, decoherence provides an elegant and minimal explanation of the quantum measurement process. To fully appreciate the importance of Zeh's work, though, it is of crucial importance to debunk a common misconception about decoherence. In contrast to how it is often portrayed, decoherence does not clarify how the alleged wave function collapse proceeds. Neither does decoherence explain how, during an interaction, a microscopic quantum mechanical wave turns into a particle, a quasi-classical object with a defined location. Decoherence does not describe what happens to the universe during an interaction or a measurement; on the contrary, it describes how an entirely quantum mechanical universe looks to a local observer.

Max Tegmark has characterized these two perspectives vividly, as the bird and frog perspectives, yet another incarnation of the film-roll and on-screen realities of the Hollywood movie plot interpretation of cosmic history. As Tegmark explains, "The theory becomes easier to grasp when one distinguishes between two ways of viewing a physical theory: the outside view of a physicist studying its mathematical equations, like a bird surveying a landscape from high above it, and the inside view of an observer living in the world described by the equations, like a frog living in the landscape surveyed by the bird." According to Tegmark, "From the bird perspective, [Everett's] multiverse is simple. There is only one wave function. It evolves smoothly and deterministically over time without any kind of splitting or parallelism. The abstract quantum world described by this evolving wave function contains within it a vast number of parallel classical story lines, continuously splitting and merging, as well as a number of quantum phenomena that lack a classical description." Quite contrary is the experience of the observers within: "From their frog perspective, observers perceive only a tiny fraction of this full reality. They can view their own . . . universe, but the process of decoherence—which mimics wave function collapse while preserving [quantum mechanics]—prevents them from seeing . . . parallel copies of themselves."[135]

Thus, decoherence has two disturbing consequences. First, the observer "splits" into "many minds"—that is, multiple copies observing each possible outcome. And second, our everyday, classical reality

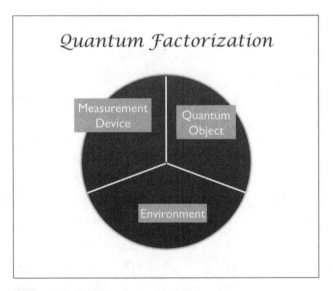

Quantum factorization: in order for a measurement to be performed, the universe has to be split into observer or measurement device, quantum object, and environment.

isn't a feature of the universe. It is a consequence of perspective. It results from the split of the universe into observer, observed, and environment (the so-called quantum factorization) plus our ignorance about the environment's exact state. It is "emergent," not fundamental. Heated debates have unfolded over the first of these points: Are Everett's parallel worlds real?

By 1967, Zeh had essentially rediscovered Everett's theory, whose papers he didn't know yet. In an unpublished draft of his famous decoherence paper, he described how "after the measurement one essentially deals with two [or more] independent worlds. It appears impossible to avoid this consequence as long as one accepts a universal validity of quantum mechanics."[136] Referring to Schrödinger's cat, Zeh arrives at the same metaphor as Everett, of a universe split "into a continuum of worlds, each of them having the cat die at another time."[137] Due to their "interaction with the environment," large, everyday objects such as cars or tables or beer bottles are "always being measured

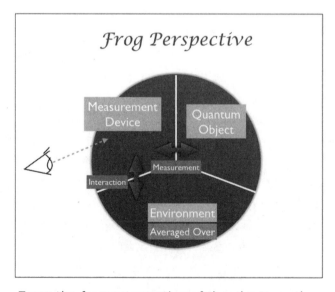

From the frog perspective of the observer, the environment is unknown and averaged over. The result is decoherence and the quantum-to-classical transition.

automatically" and thus are "automatically being separated as a consequence of the measurement."[138]

Whether the many worlds of the Everett interpretation have to be considered as "real" depends on the perspective from which this statement is made. In the bird perspective, there are no Everett branches to begin with. All that exists is a single, entangled quantum universe. From the frog perspective, only one single Everett branch is tangible (i.e., our experienced, classical universe). So what about the other Everett branches describing parallel realities? Are they real—or at least not less real than other concepts of reality we come up with, as both Everett and Zeh argued consistently? "All that we can ever know about any theory is the extent to which it seems to correspond to the real world observations we can make, and to the experiments we can do. Beyond that we never know the extent to which any of our theories capture the real reality of the universe, to the actual content of what really is out there," Everett explained to a colleague working with him at the Pentagon. "So we have no way of guessing how close any of our theories are to what may really be out there. All we can do is postulate our theoretical idea and then ask how well they correspond to experiments," Everett concludes.[139] Similarly, Zeh writes, "Since they are unobservable, the existence of those other, parallel worlds can of course be repudiated. They exist in the same sense as other heuristic fictions in physics."[140] As "heuristic fictions," Zeh identified the elements of our scientific worldview, such as, for example, quarks—the constituents of nucleons that attract each other so strongly that they are never observed individually (only the heaviest quark, the top quark discovered in 1995, decays before it melds with other quarks)—or dinosaurs. As David Deutsch emphasizes, "We don't speak of the existence of dinosaurs millions of years ago as being 'an interpretation of our best theory of fossils.' We claim that it is the explanation of fossils. And the theory isn't primarily about fossils: it's about dinosaurs."[141] Philosopher of science David Wallace agrees: "Nobody seriously believes that 'dinosaurs' are just a calculational device intended to tell us about fossils . . . And almost all of science is like this."[142] Insisting that parallel universes are "only an interpretation" instead of "a scientifically established fact" has, according to Deutsch, "the same logic as those stickers that they paste in some American biology textbooks, saying that evolution is 'only a theory.'"[143] In this sense, the "'Everett interpretation of quantum mechanics' is just quantum mechanics itself, 'interpreted' the same way we have always interpreted scientific theories in the past: as modelling the world," Wallace emphasizes, adding that the only alternative to this conclusion would be to "replace quantum theory—the most predictively powerful, most thoroughly tested, and most widely applicable theory in scientific history—with a new theory which we have not yet constructed."[144]

Yet, for neither Everett nor Zeh was the existence of parallel realities the most important point of their findings. What is often overlooked in this heated debate is that the real "elephant in the room" is the second consequence: classical reality is a consequence not only of a measurement system being coupled to an environment but also of the incomplete knowledge about this environment. This is of course a consequence of the local observer, who simply cannot have all possible information about the exact state of the entire universe. It implies that the quantum-to-classical transition is an artifact of the observer's local frog perspective, in contrast to the bird perspective, in which the entire quantum system would be observed and no quantum-to-classical transition takes place. The quantum-to-classical transition is perspectival!

Thus, in principle there are two possible kinds of quantum systems. First, there are isolated (typically microscopic) systems with no interaction with the environment. While all quantum systems we have experience with are of this type, this is naturally always an approximation. And then there is the entire quantum universe: global, encompassing, with no external environment and thus not subject to decoherence. This latter system constitutes the only true fundamental quantum state, a conjunction of opposites accommodating everything that is physically possible. The experienced world, then, emerges from this foundational "One" through decoherence. Of course, as long as we stick to the reasonable hypothesis that our consciousness is confined within our brains, there is no way we could ever experience the universe from a bird's perspective. Nevertheless, the bird's perspective isn't entirely inaccessible: as Zeh emphasized, although the frog is not able to fly like a bird and experience this fundamental reality, with "imagination guided by reason," the frog is capable of developing a description of the bird perspective of quantum reality (by solving the Schrödinger equation).[145]

As a result, whatever is achieved by decoherence, including the quantum-to-classical transition, probably the emergence of matter and possibly even space and time itself isn't a real process in the fundamental quantum universe. It only describes the impression an observer located in space and time gets about this fundamental reality. Tegmark describes this view quite fittingly as the "Platonic Paradigm: The bird perspective . . . is physically real, and the frog perspective and all the human language we use to describe it is merely a useful approximation for describing our subjective perceptions."[146] As Zeh concludes, "Quantum theory requires quantum cosmology."[147]

* * *

Decoherence marks the last step to fully establishing a monistic worldview strikingly similar to ancient beliefs. In keeping with how the Egyptians pictured their goddess Isis, the symbol of a hidden, all-encompassing unity, as veiled so that she wouldn't be exposed to the eyes of the mortals, the limited information that "frogs" can gather about the universe gives rise to decoherence and has them experience the entangled quantum universe as many, individual objects. On the most fundamental level, as observed from the bird perspective, all is One.

Yet these monistic implications are still far from becoming a general consensus. There appears to be a powerful psychological barrier that prevents us from accepting this straightforward conclusion, starting from Bohr and Heisenberg to most physicists working with quantum mechanics today. It is an inhibition that is deeply rooted in the history of science and Western religion.

4

THE STRUGGLE FOR ONE

IF EVERYTHING IN THE UNIVERSE IS MERGED INTO A single One by entanglement, if decoherence explains how this hidden unity unfolds into the planets, pebbles, and critters populating our universe, and if its profound, bizarre, and flat-out revolutionary implications are evident in the equations of quantum mechanics, we may have resolved the contradictions inherent in our universe, but another fundamental question still remains. Namely, how is it possible that such a revolutionary notion could have been ignored for so long?

We have seen how the Copenhagen physicists shied away from exploring what quantum mechanics actually means for nature and how they reclassified the foundations of physics as religion. At the same time, anyone who dared to question the Copenhagen orthodoxy and tried to find out what lay behind our daily-life, on-screen reality, whether it was Hugh Everett, H. Dieter Zeh, or John Bell, was deemed a heretic. Yet, from a historical perspective, the renegade physicists were lucky that they were merely ignored—that they were not burned alive, tortured, or killed, like Giordano Bruno and many other monistic scholars ever since the decline of antiquity. The idea of a monistic "One" was no novel, abstract scientific concept—it was a three-thousand-year-old battleground—a battleground on which Christian religion successfully claimed exclusive rights for the big picture and where science got confined to filling out the details and producing recipes for problem solving. To appreciate how deeply rooted both monism and its refusal are in Western culture, we must look back to the muddled history and origins of monism, science, and monotheistic religion itself.

Religion as Rebellion

In contrast to the monotheism of Christians, Muslims, and Jews, primal religions were typically polytheistic. Believers worshiped a pantheon of many gods, representing the diversity of the world. Yet these early religions often had a distinctly monistic flavor in which the different deities could represent the various facets of a single, unified reality. "Polytheism is cosmotheism," writes Egyptologist Jan Assmann, which is but another word for pantheism, the worship of the universe.[1] This is effectively a religious expression of monism: "the religion of an immanent god and a veiled truth that shows and conceals itself in a thousand images that illuminate and complement, rather than logically exclude, one another."[2] In cosmotheism, Assmann explains, "the principle of plurality is ineradicably inscribed into this worldview," and "the divine cannot be divorced from the world."[3]

In contrast, a distinctive feature of Judeo-Christian religion is that it regards God as a power that governs the world from outside. Thus, despite their sounding similar, "monism" and "monotheism" refer to dissimilar worldviews. In fact, according to Assmann, the monotheistic

In monism, "all is One": everything is integrated into an all-encompassing whole.

religions such as Judaism and Christianity originally developed as "counterreligions" for the marginalized and suppressed against a prevailing pagan culture: as the rebellion of an enslaved nation led by Moses out of Egypt and as an advocate for the poor and proletarian in ancient Rome. Unsurprisingly, then, these suppressed early Jews and Christians saw the world not as itself divine but as in need of divine intervention. Whereas in polytheism "the divine cannot be divorced from the world," Assmann writes, monotheism "sets out to do just that. The divine is emancipated from its symbiotic attachment to the cosmos."[4] As Assmann explains, it is a defining feature of monotheism to "release its people from the constraints of this world by binding them to the otherworldly order," to establish a "distinction between god and the world."[5] It is this quality as a counterreligion that makes monotheistic religions "aware of themselves as religions, not just in opposition to magic, superstition, idolatry, and other forms of 'false' religion, but also in contrast to science, art, politics," writes Assmann.[6]

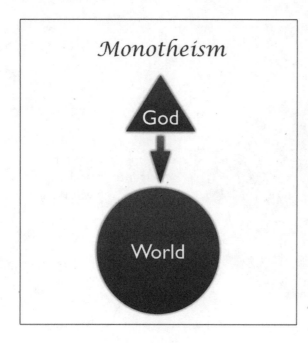

In monotheism, God is understood as different from the world that he governs from outside.

The intolerance and iconoclasm inherent in monotheism can be traced back to this origin. "The lamb is sacrificed because it corresponds to the most sacred animal of the Egyptians, the ram," and "Egypt's most conspicuous practice, the worship of images, came to be regarded as the greatest sin," Assmann details.[7] Thus it is a distinct feature of monotheistic religions that "the truth to be proclaimed comes with an enemy to be fought," Assmann writes. "Only they know of heretics and pagans, false doctrine, sects, superstition, idolatry, . . . heresy, and whatever other terms have been coined to designate what they denounce, persecute and proscribe as manifestations of untruth."[8]

This trend intensified in late antiquity with the rise of Christianity: "By worshipping the one true god, the Jews isolate themselves from the peoples, who are of no further interest to them. Through their strict adherence to the laws, they cultivate a life-form in which this voluntary isolation finds symbolic expression," observes Assmann. Christianity, in contrast, "made it its mission to put an end to this self-imposed isolation and open itself to all peoples. Now everything and everyone is excluded that refuses to take up this invitation. Monotheism thereby became invasive, at the very least, and occasionally aggressive as well."[9] It is this equally exclusive and invasive character of Christianity that became the root of future conflicts, in the relationship both with other religions and also with monistic philosophy and science.

The Genes of Science

It is common knowledge that modern science carries the DNA of ancient Greek philosophy. Less known is that this origin infused science with monistic thought. Quoting John Burnet's *Early Greek Philosophy*,[10] Erwin Schrödinger wrote, "Science can be correctly characterized as reflecting on the Universe in a Greek way."[11] The Greeks, Schrödinger went on, pioneered the split of subject and object that appeared so characteristic and necessary for any kind of objective reasoning. To Greek philosophy, it seems, monism is as alien as it can get. But this is only half of the story.

When Greek philosophy came into being, science and religion were not easily separated. In the telling of English philosopher Jonathan Barnes, the starting point of Greek philosophy as we know it was in 585 BCE in Ionia, today the west coast of Turkey, when Thales of Miletus predicted a solar eclipse.[12] Thales and most other early Greek philosophers originally pursued a more modest "substance monism." They admitted the existence of many things in the universe but thought everything was made of the same stuff or building blocks (in Thales's case, it was water)—a view akin to the concept of grand unification in modern particle physics, which tries to establish that a single quantum field (or a small set of fields) is responsible for all matter in the universe, or to string theory, which understands the various particles constituting the matter observed in nature as different oscillation patterns of fundamental strings.

Yet, at the same time that Thales and his successors laid the foundation for analytic reasoning and modern science, the more radical notion that "all is One" was also promulgated elsewhere. It spread through literature and various esoteric sects and cults, some of them influenced or imported from Egypt. These spiritual gatherings existed in parallel with the official religion devoted to the famous polytheistic Olympian pantheon around Zeus and Hera. One of the earliest pieces of Greek poetry, *Works and Days* by the eighth-century-BCE writer Hesiod, describes a prehistoric paradise with obvious parallels to the biblical Garden of Eden. In this "Golden Age," humanity was imagined to live in harmony with nature, often depicted as dancing with nymphs to the flute-playing of Pan, the Greek god of nature. Centuries later, believers in "Orphism" composed hymns to night and heaven, wind and fire, sea and earth, moon, stars, and planets, and an all-encompassing nature:

> *Nature, all parent, ancient, and divine, O much-mechanic mother, art is thine;*
>
> *Heav'nly, abundant, venerable queen, In ev'ry part of thy dominions seen.*[13]

Another hymn refers directly to a monistic whole, allegorized by the nature god Pan, whose name in Greek also translates as "all":

> *I call strong Pan, the substance of the whole, Etherial, marine, earthly, general soul,*
>
> *Immortal fire; for all the world is thine, And all are parts of thee, O pow'r divine.*[14]

The group traced its roots back to the mythical poet Orpheus, who, it was said, could charm all men and women, animals, and even stones with his wonderful music. According to legend, his music also allowed him to enchant the gods of the underworld to let him at least temporarily retrieve his wife, Euridike, from death.

A similar myth about death and rebirth and the cycles of life served as a central theme for one of the most famous and influential mystery cults of antiquity. The Eleusinian mysteries promoted a primeval, cyclic grasp of history and time, also found in Egyptian tradition. In contrast to the modern conception assuming a linear succession of past, present, and future, the Egyptian concept of time was inspired by the recurrent annual seasons, the periodicity of planetary orbits, and the tides of the river Nile. The Greeks adapted this concept in the myth of Demeter, the goddess of the harvest and agriculture, and her daughter Persephone, who got abducted by Hades, the god of death, to make her his bride in the underworld. When Demeter learned about what had happened, the grieving goddess stopped the growth of fruits, grains, and vegetables, and as a consequence humans, animals, and gods were starving. At some point Zeus, the king of gods, intervened and negotiated a compromise. According to this settlement, Persephone had to spend one-third of the year as queen of the dead with Hades. This period became the winter, when Demeter grieved and the fields withered, but for two-thirds of the year, Persephone could live with her rejoicing mother, allowing nature to flourish during the spring and summer seasons.

The popularity of this and other mystery cults in ancient Greece has been explained as an expression of humanity's innate longing for nature. While antiquity may feel like a long, long time ago to us, even in the midst of ancient Greek society, humans responded to the rise of the city-state with nostalgia for a lost time that came before. Since prehistoric times, together with the organization of societies, the role of religion had changed. The specialization of labor had led to an alienation from nature, and the growing societies were in need of ethical codes to replace personal acquaintance as a foundation of social harmony. Religion changed from a body of primitive, proto-scientific explanations into a set of moral laws. Thus, while the official religion had evolved away from monism when the small prehistoric bands and tribes progressed into chiefdoms and states, these cults imparted a lost sense of integration into the natural cycles of life, of "being one with nature," to the burgeoning ancient civilizations. Moreover, with their secrecy and initiation rites, these cults revived the familiarity that came along with living in a small tribal society.

At the same time as Lao-tzu wrote the *Tao Te Ching* in China, these different threads started to converge. Only a couple of years after the solar eclipse predicted by Thales was observed, two philosophers were born on the Ionian coast, a hub for trade with Egypt and the Middle East. Both men later emigrated to Greek colonies in today's southern Italy, where each of them founded an influential school. Pythagoras, famous for his alleged discovery of the geometrical relation among the three sides of a right triangle, left his native Greek island of Samos, a couple of miles off the Ionian shore, and became the founding father of a close-knit group of mathematician-philosophers. Scholars are undecided whether Pythagoras was a rational mathematician or a shamanic figure—or something in between. Several of Pythagoras's ancient biographers reported that Pythagoras spent some time in Egypt in his youth, where he learned from the Egyptian priests and adopted the practice of secretiveness. Ion of Chios, a scholar born shortly after Pythagoras's death, suggested that Pythagoras was actually the author of the Orphic hymns. The fact is, Pythagoreans pioneered the mathematical and experimental study of music; discovered the relations between the ratios of small natural numbers, the length of the strings of a lyre, and the tone pitch; and evolved into a close-knit religious group that believed the universe was governed by math and harmonies. Within this numerological philosophy, Pythagoras's successor, Philolaus, identified the number one as the center of the universe,[15] and the first-century-CE philosopher Eudorus later testified, "The Pythagoreans teach that on the highest account the One is the principle of all things . . . [R]anked below . . . are all things that are conceived in terms of opposition . . . Matter and all Beings have come into Being from it."[16]

Pythagoras's contemporary, Xenophanes, came from the Ionian city of Colophon and resettled in Elea, an ancient port on the Tyrrhenian coast some ninety miles south of Naples. Most famous among his followers became his student Parmenides. As far as we know, Parmenides composed only a single work, a poem in epic hexameter titled "On Nature" that narrates the author's mystical journey to the home of the gods, culminating in the revelation of cosmological and philosophical truths by an unnamed goddess. Parmenides describes how the goddess reveals to him two means of inquiry. Around the same time when Heraclitus back in Ionia proclaimed, "From all things One and from One all things" and "Nature loves to hide," and deposited his book at the feet of the local incarnation of Isis, the statue of Artemis in her great temple, one of the seven wonders of the ancient world in the regional metropolis Ephesus, Parmenides wrote about "the unmoved heart of persuasive reality,"[17] described as "the one, that is and that is not not to be" that "attends upon true reality."[18] In contrast to Heraclitus, who emphasized the vicissitudes of nature, Parmenides stresses that this underlying, true reality is timeless and eternal, just like the eternal principles governing the change of seasons celebrated in the Eleusinian mysteries. And just like moviegoers ignorant of what is going on in the projector room, Parmenides emphasizes that "mortals," portrayed as "deaf and blind at once, bedazzled, undiscriminating hordes," "know nothing" about this fundamental reality.[19] While, as a consequence of the fragmentary record and the poetic language, several competing interpretations of Parmenides's poem exist, a common and influential reading is that Parmenides provides an early comprehensive exposition of the seemingly paradoxical view that "there exists exactly one thing,"[20] "ungenerated and imperishable, entire, unique, unmoved and perfect" and composed of complementary principles such as light and darkness.[21]

In fact, a significant part of the Greek philosophical tradition carries on this distinctively monistic flavor—most prominently Platonism. It was through Platonism that Orphism and the mystery cults, Pythagoras and Parmenides were blended into a powerful narrative that engendered science, religion, and the entire history of philosophy ever since.

Plato's Secret

"The safest general characterization of the European philosophical tradition is that it consists of a series of footnotes to Plato," wrote Alfred North Whitehead, an English mathematician and philosopher of the nineteenth and twentieth centuries.[22] Plato was a prime representative of the fifth-century-BCE "Golden Age of Athens," a time that encompassed the building of the Athenian Acropolis, Hippocrates and Herodotus becoming the fathers of medicine and history, Phidias creating his famous gold-plated marble statues, and dramatic poets like Aeschylus, Sophocles, Euripides, and Aristophanes composing their immortal plays. A student of Socrates and teacher of Aristotle, Plato arguably became the most influential philosopher in history. His school, "the Academy," has been characterized as the first academic institution in the Western Hemisphere, which, with some breaks and reestablishments, continued to shape the ancient world for more than nine hundred years. Even if monism appears to be a global phenomenon, it was Platonism—particularly its history and changeful interactions with Christianity—that influenced the Copenhagen physicists' reception of the quantum world.

Plato was an initiate of Eleusis and a close friend of Archytas, one of Pythagoras's most prominent successors, who once saved his life. His works *Philolaus* and *Timaeus*, the book that later inspired both Johannes Kepler and Werner Heisenberg, are strongly informed by Pythagorean ideas, and in his book *Parmenides*, he meticulously analyzes the philosophy of the Eleatics. Accordingly, monism became a trademark of his school. A particularly vocal champion of the monistic trend in Platonism was the third-century-CE Neoplatonist Plotinus, who wrote in his magnus opus, *The Enneads*, "The One is all things and no one of them; the source of all things is not all things; and yet it is all things in a transcendental sense—all things, so to speak, having run back to it: or, more correctly, not all as yet are within it, they will be." Plotinus also addresses the question of how this primeval One is related to the plurality we observe all around us: "It is precisely because there is nothing within the One that all things are from it: in order that Being may be brought about, the source must be no Being but Being's generator, in what is to be thought of as the primal act of generation."[23]

Yet, in contrast to his followers, Plato himself is not famous for his monism. Plato's philosophy is perhaps best known for his "theory of ideas" or "forms" underlying each element of visible experience, just as the different features on the film roll underlie the on-screen reality, a doctrine that he fleshed out in his early dialogues. In fact, in order to find a prominent credo of monism within most of Plato's written work, one has to read between the lines. Some scholars have even argued that Plato had been misunderstood or deliberately misrepresented by his Platonist, monist followers. But then, Plato also was skeptical about the power of the written word: "There neither is nor ever will be a treatise of mine on the subject. For it does not admit of exposition like other branches of knowledge," the philosopher confessed in his *Seventh Letter*.[24] Ever since, it has been speculated whether Plato had reserved core ideas of his philosophy for a secret, exclusively oral dissemination. Aristotle and others gave account of such an "unwritten doctrine" taught at Plato's Academy. When, in the twentieth century, philosophers in Tübingen and Milan tried to reconstruct this unwritten lore, they concluded, "Plato's philosophy focuses exclusively on 'The One'"[25]—an all-encompassing reality that became prominent later in the teachings of Neoplatonists but rarely appears explicitly in Plato's own written work.

One notable exception is his puzzling book *Parmenides*, in which he discussed the Eleatics' claim that "all is One." Atypically, Socrates, who usually represents Plato's voice, appears here as a young disciple of an older Parmenides. Next, the dialogue seems to criticize some aspects of Plato's own earlier teachings about the variety of ideas and forms underlying experience. Finally, the narrative ends abruptly, having mainly exposed the paradoxes associated with its subject instead of its benefits. For example, since an all-encompassing One embraces both things that are blue and other things that aren't, it can't be either entirely blue or not-blue itself. Nor can it be composed of things being blue and not-blue, since in the latter case it would be many things; it wouldn't be "one" anymore. Plato used different characteristics, but the logic remains the same: "the one can neither be the same, nor other, either in relation to itself or other."[26]

Other monistic philosophies struggled with the same paradoxes: if everything is integrated into a single concept, this concept necessarily becomes a conjunction of opposites. It defies any description in terms of concrete properties, as highlighted, for example, in the first sentences of Lao-tzu's *Tao Te Ching*: "The Tao that can be spoken of is not the constant [true] Tao; The name that can be named is not the constant [true] name."[27]

These intricacies are, in fact, precisely what Niels Bohr described with his notion of complementarity. Just replace "blue" and "not-blue" with "particle" and "wave," and Plato's *Parmenides* reads like a standard textbook introduction to quantum mechanics. As Carl Friedrich von Weizsäcker wrote, "We find . . . the foundation of complementarity already foretold in Plato's Parmenides."[28]

It is important to stress, though, that Parmenides and Plato didn't anticipate Bohr's denial of a reality beyond the experienced world. For the Greek philosophers, a sharp separation of science, philosophy, and religion didn't exist. Thales of Miletus, for example, understood the universe to be a living organism whose soul could be identified with God. Parmenides's poem was known to the ancients as *On Nature*, his account of a heavenly journey on which a goddess revealed to him as fundamental truth a unique, monistic, unchanging, and timeless reality that contrasts with the apparent world's illusion of a becoming, developing cosmos. Likewise, in Plato's book *Timaeus*, the physical world is understood as an imitation of the eternal world by the "demiurge," a divine craftsman. Accordingly, Plato's One is "transcendental" and "metaphysical." It isn't directly observable and lies beyond the domain of everyday physics. Yet, to Plato, metaphysics wasn't unreal. In fact, it was more real than the observable phenomena; it was the real world rather than its shadow. Eventually, the effect of how Platonic philosophy got appropriated by Christianity pushed Plato's monism into an otherworldly realm.

From One to God

The Areopagus is an assembly of rocks some two hundred meters northwest of the Athenian Acropolis. In contemporary times, on warm summer nights the place is crowded with a colorful blend of students and young tourists from all over Europe and the rest of the world, playing guitar, drinking cheap wine, and peering magnetized at the illuminated, ancient citadel of the pagan temple to the city's namesake, the Greek goddess Pallas Athena. As described in the Bible, almost two thousand years ago, Paul the Apostle climbed this rock to speak to the Greeks. Considered to be one of the most important figures in early Christianity, Paul was instrumental in opening the new religion to non-Jewish believers and thereby instituting its role as a future world religion. And in order to do so, Paul started to integrate pantheistic, Greek philosophy into the originally Jewish body of Christian faith: "He is Lord of heaven and earth, dwelleth not in temples made with hands . . . For in him we live, and move, and have our being," says Paul in Acts of the Apostles.[29]

The Areopagus sermon provides a prime example of the efforts of early Christianity to proselytize to the gentiles and spread the gospel to all nations. Still, for more than three centuries after Paul's speech, Platonism competed with Christianity to become the dominant worldview in the Mediterranean world. Christianity finally prevailed in these struggles but inherited Platonic ideas in their reincarnation as "pantheism" or "panentheism," where the universe was identified totally or at least partly with God.

One famous example is the Christian doctrine of the Trinity of the Father, the Son, and the Holy Spirit. This crucial concept, which makes the divinity of Jesus Christ possible in the first place, originally goes back to the Platonist Porphyry of Tyre, a strict adversary of the

Christians, who understood the Trinity as a conjunction of "being," "life," and "spirit."[30] An early champion of this trend to integrate Platonic philosophy into Abrahamic faith was the Jewish philosopher Philo of Alexandria, who lived around the time of the birth of Christ and described God as "unnamable," "unspeakable," and "incomprehensible," alluding to the veiled Isis as a symbol of a hidden, monistic One.[31] Philo and his Christian followers Gregory of Nyssa and Dionysius the Areopagite then compared the ascent in Plato's Allegory of the Cave with Moses's journey up to God on Mount Sinai.[32] According to philosopher Charles H. Kahn, "Philo's great achievement was to make use of Greek philosophy, and the Greek allegorical technique . . . in order to provide a systematically philosophical reading of the Hebrew Bible."[33] In this context, Kahn wonders, though, how much this monotheistic reading also changed the interpretation of Platonism itself, about "how far the new perspective introduced by Philo from Jewish monotheism contributed significantly to the increasingly transcendental conception of deity in . . . Neoplatonic philosophy."[34] In fact, Philo's philosophy, describing a hierarchical pattern of powers between God and the world, leads in effect to an increase of "the distance between the highest deity (or the one God) and the natural world," as Kahn observes.[35]

Apart from confounding Platonism in the process of appropriating it, Christianity didn't draw on monism alone. "Manichaeism," named after its Persian prophet Mani, advocates a worldview quite opposed to monism, believing that the world is caught in an epic struggle between good and evil. Through Manichaeism, "dualistic" concepts such as angels and demons, God and devil, and paradise and hell received their prominent role among Christian beliefs. A pivotal figure in this process was the bishop, philosopher, and saint Augustine of Hippo, one of the four major fathers of the Catholic Church. Augustine was born in 354 CE as the son of a Christian mother and a pagan father in the Roman province Numidia, now Algeria. As a young man he first became a Manichaean, before he moved to Rome, fell under the influence of Platonism, and eventually converted to Christianity. "In the Platonists, God and His Word are everywhere implied," writes Augustine, who describes how in their books he found "though not in the very words yet the thing itself and proved by all sorts of reasons."[36] Indeed, Augustine's early conceptions of truth, time, and love exhibit a distinct Platonic flavor. In his later works, however, dualist notions become increasingly important, including the idea of humanity as a "mass of perdition" and of a just, holy, and universal war between the "City of God" and adversary powers aligned with the devil; they also convey a distinct anti-Judaism and antipaganism.

The combination of these notions with an intensified disdain for physical nature in favor of an otherworldly beyond drove Christian zealots to despise earthly pleasures, destroy ancient temples, burn invaluable books, and enforce repressive sexual mores. "During the fourth and fifth centuries, the Christian Church demolished, vandalized and melted down a simply staggering quantity of art. Classical statues were knocked from their plinths, defaced, defiled and torn limb from limb. Temples were razed to their foundations and burned to the ground," as Catherine Nixey describes in her revealing account of how the new

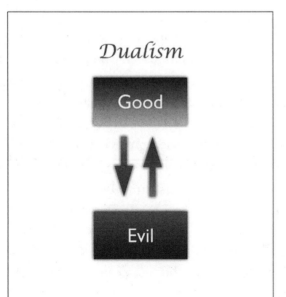

In dualism, the universe is perceived to be governed by opposing powers such as "good" and "evil."

world religion contributed to the decline of antiquity.[37] The scorn of everything mundane didn't stop when it came to intrusions of privacy: "Everything from the food on one's plate (which should be plain and certainly not involve spices), through to what one got up to in bed (which should be likewise plain, and unspicy) began, for the first time, to come under the control of religion," Nixey writes, providing concrete examples: "Male homosexuality was out-lawed; hair-plucking was despised, as too were make-up, music, suggestive dancing, rich food, purple bedsheets, silk clothes . . . The list went on."[38] Instead, Christian saints glorified extreme asceticism, self-flagellation, and martyrdom. Nixey pictures how Saint Antony scorned his body and "assaulted it on a daily basis, refusing to use oils to anoint and clean it, and instead wore a hair shirt and never washed," how he "was burning with a desire for martyrdom," and how Syrian monks "for the love of God, lived out their entire lives standing on pillars, or in trees, or in cages."[39]

Naturally, this zeal provoked conflict. A famous victim of the increasing tensions between Christians, Jews, and pagans was the astronomer, mathematician, and philosopher Hypatia of Alexandria (350–415). The city of Alexandria, praised as "the flower of all the towns," had been founded in 331 BCE by Alexander the Great as a Greek city on Egyptian soil.[40] More than being the second-biggest city in the Roman Empire and home to one of the seven wonders of the ancient world, the lighthouse Pharos, Alexandria also was an illustrious center of learning. The city's intellectuals included Archimedes, Euclid, Aristarchus of Samos, who proposed the first heliocentric model, and Galen, probably the most accomplished physician of antiquity. Alexandria's intellectual heart was the Musaeum, an academy forged after the exemplar of Plato's Academy in Athens, which hosted some of the best scholars in the empire and probably the greatest library in the ancient world, holding many tens of thousands of scrolls. Hypatia's father, the mathematician Theon of Alexandria, was a member of the Musaeum and authored an edition of Euclid's *Elements* that, until the eighteenth century, was the only known version of the mathematical treatise that has been praised as the "most influential textbook of all times."[41] Hypatia herself was one of the first female mathematicians known and is remembered as the last great Alexandrian astronomer. According to a contemporary historiographer, she taught philosophy in the tradition of Plato and Plotinus and attracted large audiences, "many of whom came from a distance to enjoy her learning."[42]

As a local celebrity, Hypatia got drawn into an escalating conflict between Christians and Jews. When Cyril, the bishop of Alexandria, instigated violent mobs to plunder and ravage synagogues and Jewish quarters, Hypatia's friend, the moderate Christian aristocratic governor Orestes, complained to the Roman emperor about Cyril's actions. This only made Cyril more furious, and soon the governor himself was attacked by a group of rioters and hit with a stone. After Orestes had the perpetrator captured and tortured, popular rage focused on Hypatia. As a wealthy and educated woman and a member of the pagan minority, the famous philosopher must have appeared suspicious to the poor, uneducated, and fanatic rioters, and soon rumors spread that she was behind Orestes's actions. One day, as Hypatia rode through the streets of the city, she ran into an ambush: "They shredded her clothes and her body with pottery fragments, tore out her eyes, dragged her corpse through the streets of Alexandria, and then burned her remains."[43] While most people in the empire, both Christians and non-Christians, were appalled by this unprovoked murder, the incident ended the career of Orestes, though Cyril remained an influential figure in the church and was finally sainted. The repression of pagan religion and ancient philosophy only became more intense.

As Christianity finally took over as the official state religion of the Roman Empire in the fourth century, this development put the official church in a quandary. It was now torn between the wealth and power that would come with its newfound dominance and the enduring disdain of its fundamentalist believers for worldly perquisites. One strategy for reconciling these opposing trends was to turn on monism itself and to condemn it for its secularization of God. In the centuries after, the ramifications of this conflict would influence religion, religion would shape philosophy, and philosophy would harm science. In 356 CE the death penalty was instituted for those who made sacrifices to the pagan gods. In 388 CE arguing about religion was outlawed. And in 407 CE the old pagan festivals were forbidden. The monistic tradition was gradually choked off.

Light in the Orient, Dawn in the West

As Christianity rose, the Roman Empire declined, at least in its central and western provinces around Italy. When Constantine the Great transferred the imperial residence to Constantinople, he accelerated the split of the Mediterranean world into East and West. As a consequence Greece's language, philosophy, and books were forgotten in western Europe. Then, in 410 and 455 CE the Visigoths and Vandals were sacking Rome, and in 476 CE the Germanic military leader Odoacer dethroned the last West Roman emperor before he got murdered himself by Theodoric the Great, the king of the Ostrogoths. Theodoric's chancellor, Boethius, can be considered antiquity's last Platonist. When he tried to revitalize the relations between the Western and Eastern Roman Empires, he was accused of treason and sentenced to death. While in prison awaiting his execution, he wrote *The Consolation of Philosophy*, a fictitious dialogue with a beautiful woman personifying philosophy who helps him to accept his fate. The work became instrumental in infusing medieval thinking with the notion of "the One," while many other sources ran dry. Boethius was executed in 524. Five years later—in 529—Plato's Academy in Athens was finally shut down by the Eastern Roman emperor Justinian. Its remaining members first found asylum at the court of Persia's philosopher king, Khosrow I, who founded an academy in Gondishapur to have Plato and Aristotle translated into Persian. One year later the philosophers returned to the Roman Empire and probably settled down in Harran in Upper Mesopotamia. In the town near today's Turkish border with Syria, a new academy was founded that existed for four hundred years and may have inspired the Islamic Golden Age with monistic philosophy.[44]

In the Eastern Roman or Byzantine Empire and the Islamic world, monistic philosophy and antiquity's knowledge were preserved for posterity. Starting from the seventh century, the Islamic expansion successively conquered Syria, Egypt, Persia, and Northwest Africa and invaded the Iberian Peninsula until it reached its limit in western Europe, stopped by Karl Martell in 732 CE. Yet, unlike the Christians, the Muslims preserved, revived, and developed much of the Greek science, philosophy, and arts that they found in their newly claimed territories. In the eighth century the Abbasid caliphs established the House of Wisdom in Baghdad, a research and learning center equipped with a library based on the model of Plato's Academy in Athens or Alexandria's Musaeum. Similar institutions, such as the House of Knowledge in Cairo (1004 CE) and the Houses of Wisdom in Cordoba and Seville, followed.

Some of these places became strongholds of monistic philosophy. A prominent figure in Baghdad's House of Wisdom, for example, was the philosopher, mathematician, physician, and musician Abu Yusuf al-Kindi (801–873 CE), who became known in the West as Alkindus, or simply as the Philosopher of the Arabs.[45] Al-Kindi got appointed by several caliphs to oversee the translation of Greek books into Arabic, and around 840 CE he adopted major parts of Plotinus's *Enneads* in his books. In his treatise "On First Philosophy," he writes that "oneness is in all . . . things" and concludes that thus there necessarily exists "a true One whose unity is not an effect."[46] Even more famous than al-Kindi was Abu Ibn Sina (980–1037 CE), who was born in central Asia (today's Uzbekistan), served later at different Persian courts, and became known in the West as Avicenna, or the Prince of Physicians.[47] According to Jon McGinnis, Neoplatonic elements featured prominently both in Ibn Sina's philosophy and in the writings of Arabic medieval philosophers more generally: "The Neoplatonic One . . . is the principle . . . of all unity in the cosmos" in such a way that "the very existence or being of the cosmos overflows or emanates from the One."[48] The monistic tradition in Islam is also evident in its mystic variety known as Sufism, revealed, for example, by the beautiful verses in which the poet Jalal al-Din Rumi (1207–1273) describes God: "My place is the Placeless, my trace is the Traceless; . . . I have put duality away, I have seen that the two worlds are one; One I seek, One I know, One I see, One I call."[49]

In western Europe, however, most of this tradition was lost. The ancient knowledge—including monistic philosophy—only slowly transfused back when relationships between western Europe and the Muslim empires intensified in the thirteenth century, through both increasing commerce and the military recaptures of Sicily by the Normans, Palestine by the Crusaders, and the Iberian Peninsula during the Reconquista. This entailed consequences for how the ancient philosophy was conceived of and valued: monism was deified or demonized, but almost always pushed into an otherworldly realm. To illustrate the paradoxical dealings of Christianity with monism, it is instructive to compare the trajectories of five philosophers: Dionysius the Areopagite, John Scotus Eriugena, Meister Eckhart, Nicholas of Cusa, and Giordano Bruno. All five of them draw heavily on Platonism and identified Plato's "One" with the Christian God, and all of them struggled to discriminate God from the natural world. But although their philosophies exhibit striking similarities, their personal lives and fates couldn't have been more different.

The Anonymous Philosopher and His Unknown God

Little is known about the first in this group. The man who became famous as Dionysius the Areopagite lived during the late fifth and early sixth centuries amid the often chaotic transition from late antiquity to the Middle Ages. We know so little about him primarily as a consequence of his clever survival strategy. Most probably a Syrian monk, he wrote under the name of a convert mentioned in the Bible's account of St. Paul's Areopagus sermon, a forgery not uncovered until the turn of the twentieth century. Under this nom de plume, the alleged Dionysius ventured on to blend Platonism into Christianity. As philosopher of religion Deirdre Carabine has pointed out, this deception served three different purposes at once. First, "the assumed identity of St. Paul's Athenian convert was not arbitrarily, but rather well chosen, for it heralds the meeting of Athens and Rome at the altar to the unknown God."[50] It thus emphasized the core concern of Dionysius's philosophy: to reconcile Christian religion with Greek philosophy. Next, the trick served as a publicity ploy: "Without the authority of the sub-apostolic status," Carabine writes, "his works undoubtedly would not have exerted the enormous influence they did upon the philosophical and theological development of Christian thought."[51] Finally, "in perpetrating one of the greatest forgeries in early medieval times, Dionysius undoubtedly spared himself the indignity of condemnation and ensured the survival of a method of theological analysis"—not to mention his monistic philosophy.[52]

Exploiting the authority and protection his false identity granted him, Dionysius boldly identified the monistic One with God. "The name 'One' means that God is uniquely all things . . . and that he is the cause of all without ever departing from that oneness. Nothing in the world lacks its share of the One . . . so everything, and every part of everything, participates in the One. By being one, it is all things," the philosopher wrote in his book *The Divine Names*.[53] Among all possible names for God, Dionysius determined, "One" is "the most enduring of them all."[54] Just like Philo of Alexandria's God, Dionysius's One, aka God, is "unspeakable," "unknowable," and "beyond the reach of mind or of reason," yet not totally divorced from the world either: "We know him from the arrangement of everything, because everything is, in a sense, projected out from him . . . God is therefore known in all things and as distinct from all things," Dionysius wrote.[55] Dionysius's idea of a reality projected out of God is strikingly reminiscent of the projector realities that help explain quantum mechanics. This raises the same question that John Bell posed in his criticism of the Copenhagen interpretation: "How exactly is the world to be divided into speakable apparatus . . . that we can talk about . . . and unspeakable quantum system that we can not talk about?"[56]

Following Philo of Alexandria, Dionysius tried to harmonize religion and reason by establishing "the Unknown God" as a transcendent, unknowable, and unspeakable unity beyond the tangible world: "There is no speaking of it, nor name nor knowledge of it. Darkness and light, error and truth—it is none of these. It is beyond assertion and denial," he wrote.[57] As Carabine emphasizes, "Few later Christian writers of the medieval period take negation so seriously or apply it in such a radical fashion."[58] On the positive side, Dionysius managed successfully to preserve monistic philosophy for the Western tradition. For many centuries to come, Dionysius's work remained the only unsuspicious source of monistic thought in western Europe, and his "negative theology" was able to define a clear separation between science and religion that helped to avoid conflict. On the negative side, by adhering strictly to the separation between God and world that got so deeply engrained in Christian thought, he eventually contributed to the dynamic that kept Bohr and Heisenberg from accepting a fundamental quantum reality fourteen centuries later.

According to the unspoken set of rules that Dionysius helped bolster, any effort to understand the foundations of reality within the framework of science was discouraged. Any breach into the territory of this unknown God had to be understood as a trespass into the realm of religion and consequently as a threat to the fragile peace between science and religion. In fact, this notion still determines the terminology of philosophical subdisciplines to the present day: "Metaphysics," that is, what lies beyond physics, includes the foundations of physics as the philosophical thought about God. In practice, everything "too metaphysical" is often met by physicists with skepticism or derision.

Yet, by and large, Dionysius's forgery paid off, and his philosophy struck a nerve. As Deirdre Carabine writes, his work had "enormous repercussions in both Eastern and Western Christian scholarship throughout the Middle Ages and indeed right down to the present day."[59] As an ultimate honor, Dionysius was considered one of the "church fathers," the scholars who established the intellectual and doctrinal foundations of Christianity.

A Revolutionist in Paradise

Around three hundred years later, starting from the eighth century CE, the Carolingian kings of Francia began to make efforts to stimulate the revival of learning and support the reform of church and education now known as the Carolingian Renaissance. The Irishman John Scotus Eriugena was part of an illustrious group of scholars from the British Isles and Italy at the court of Charles the Bald, which his grandfather Charlemagne had started to gather around his leading scholar, Alcuin of York. In particular England and Ireland had been relatively unimpaired by the upheavals due to the barbarian invasions and migration period following the fall of Rome. As a consequence, the isles became famous for learning and scholarship in the early Middle Ages, and at the Carolingian court the expertise of these intellectuals was most welcome in order to copy and preserve texts of classical authors. Role models for these efforts were the Christian Roman Empire of the fourth century and Byzantium. Just like his grandfather Charlemagne, "Charles the Bald," writes Carabine, "was enamored with things Byzantine."[60] According to historian of philosophy Kurt Flasch, a veritable "book trail from Rome to York and finally Fulda" developed.[61]

In 848 AD a monk named Gottschalk stirred up trouble at the monastery in Fulda. In the midst of what is often depicted as the "Dark Ages," this episode initiated a series of events that led to an unorthodox theological interpretation of the Fall of Man that appears as strikingly modern. Trying to reconcile Plato's monism with the Christian conceptions of heaven and hell landed Christian theology in a paradox, as we shall see, but at the same time the episode provides a striking allegory for decoherence and what is known as the "quantum factorization problem," the problem that if all is One, where is the observer who assumes such an instrumental role in the emergence of the classical world coming from?

The cause of the conflict was a treatise by Gottschalk that denied the existence of free will. Relying on the authority of St. Augustine, the obnoxious monk claimed that every man and woman was predetermined by God to end up in hell or heaven, irrespective of his or her lifestyle and accomplishments. Gottschalk went as far as to advocate that God wouldn't want every man and woman to be saved and that, quite the contrary, half of humanity should be condemned to hell. Although Gottschalk's reading was indeed in accord with Augustine's later writings, from a political viewpoint, such a stance was deemed dangerous: it risked undermining the ethical basis of the medieval society based on Christian faith. Accordingly, a synod sentenced Gottschalk to be flagellated and incarcerated for the rest of his life. But this wasn't the end of the story. To underpin its verdict, the synod asked Eriugena, court grammarian of Charles the Bald, king of western Francia and later Roman emperor, to draw upon his expertise.

As desired, in his report Eriugena indeed condemned Gottschalk's teachings. He attested to Gottschalk's "perverse thinking" and criticized his treatise as a "monstrous, poisonous, deadly doctrine," but then he resolved the argument in a rather unexpected way.[62] Taking the liberty of arguing that Augustine "does not mean what he says," Eriugena drew on Platonism to develop a radical reinterpretation of the earthly and divine realities and the Christian notion of heaven, hell, and the Fall of Man—and what's more, a monistic philosophy bearing striking similarities to the workings of quantum mechanics.[63] For a start, Eriugena rebutted Gottschalk by asserting that God is so entirely good that to him any relation with evil or sin is entirely alien. Consequently, whatever was evil wasn't God's will. Instead Eriugena conceived the evil in the world as a lack of God. Likewise, God wouldn't ever condemn anybody to hell: "God does not curse the things which He made, but blesses them," Eriugena wrote.[64] Thus, rather than being a physical place in space and time, hell had to be understood as a psychological condition afflicting those neglecting their connection with God.

So far, this doesn't sound like it would have much bearing on physics. But then Charles the Bald asked Eriugena to translate a precious manuscript that his father, Louis the Pious, had obtained as a gift from the Byzantine emperor, containing the complete works of Dionysius the Areopagite. Following in the footsteps of Dionysius, Eriugena identified Plato's One with the Christian God. For a Neoplatonist philosopher this was a straightforward conclusion: as he was convinced that all is One, there was scarcely a better candidate to find. For a devout Christian of the Middle Ages, living in a universe populated by angels, demons, and hellfire, this was a radical leap. "All things come from the One, and there is nothing which does not come from It"; thus "God is all things and all things God," Eriugena writes in his opus magnum, *The Division of Nature*.[65]

As a consequence, Eriugena arrived at an appreciation of nature conspicuous for its monism. "The beauty of the whole established universe consists of a marvellous harmony of like and unlike in which the diverse ... and various species and ... substances ... are composed into an ineffable unity."[66] In other words, "there is a most general nature in which all things participate, which is created by the One Universal Principle" and "from this nature corporeal creatures are derived, ... [like] streams which, issuing from one all-providing source, ... break out ... in the different forms of the individual objects of nature"—a nature, as should be emphasized, that is governed by science and mathematics: "The infinite multitude of all things visible and invisible assumes its substance according to the rules of numbers," Eriugena explains.[67] His heroes are the "first of all philosophers," the "supreme" Pythagoras and Plato, "the greatest of those who philosophized about the world" and the only one who "discovered the Creator from the creature."[68] As summarized by Carabine, the central idea of Eriugena's book is that "creation is the manifestation of God and, therefore, is sanctified, since all things have come from the same source."[69]

What was described as the Fall of Man and the original sin in the biblical Genesis is now understood as a metaphor for the individual separating herself out of this unity with God. Even the emergence of being in space and time was attributed to this separation. As Carabine emphasizes, "Perhaps the most important consequence of the fall is that it effected the creation of the body and the material world."[70] Similar motifs would recur a thousand years later in the philosophy of

Georg Wilhelm Friedrich Hegel, Friedrich Wilhelm Joseph Schelling, and Søren Kierkegaard. Even more amazingly, Eriugena's work looks like a metaphor for decoherence, where the separation of the universe into subject, object, and environment constitutes a local perspective onto the world and in this way can lead to the emergence of time and localized, classical objects—and, of course, of good and evil, since without time nothing happens, so nothing can be considered evil. Just as in decoherence theory, where matter and potentially space and time emerge from a quantum universe or universal "One," Eriugena's philosophy suggests that, as Karl Jaspers summarized, "man has turned away from God and is now on his own. He lost his Being in God's eternity. He has to be present in certain places in space and time. Matter that once was spiritual itself is now physical. The Unity of being has been ripped apart. Everything is now divided, God and world, mind and matter, species and individual, man and woman."[71]

While this is already stunningly close to how quantum mechanics functions, it still doesn't exhaust the parallels between decoherence and Eriugena's conception. Even the bird perspective conceptualized by Max Tegmark and H. Dieter Zeh was anticipated by Eriugena when he credited St. John as the voice of the solitary eagle in high flight, "not of the bird who soars above the material air . . . orbiting the entire sensible world—but the voice of that spiritual bird who . . . transcends all vision and flies beyond all things that are and are not"[72] and who, as Carabine explains, "can see the whole of reality from its vantage point in the sky."[73] Just as entanglement unites the universe in quantum cosmology, for Eriugena it is "the pacific embrace of universal love" that "gathers all things together into the indivisible Unity which is what He Himself is, and holds them inseparably together."[74]

Revolutionary as he was, Eriugena could draw on ancient paradigms, both from philosophers whom he read or translated himself and from others he learned about indirectly through other authors. Most notable in this tradition is the *Symposium*, often named as Plato's most beautiful work. It describes a convivial gathering of a group of men in Athens, among them the philosopher Socrates, the politician

Alcibiades, and the comic playwright Aristophanes. Since most of the participants still suffer hangovers from the previous night, they decide to cut short on drinking and rather entertain themselves with a contest of speeches about "Eros," or "Love." When it is Aristophanes's turn, he tells a myth according to which in primeval times people had both four arms and four legs and two sets of sexual organs. Equipped this way these primordial humans were strong enough to rebel against the gods. In the ensuing struggle, the humans were defeated and, as a punishment, ripped apart. Ever since, humans have felt imperfect and alone, and this, according to Plato's Aristophanes, is the reason why humans seek love: to find their lost counterpart and to become one and whole again.[75]

Already Plato's narrative, which has been frequently adopted in the literature since, bears an interesting resemblance to the book of Genesis and the Fall of Man. Also in the biblical context, Eve is ripped out of Adam, and a primeval paradise is lost as a consequence of Adam and Eve's rebellion against God. In the Old Testament, the Fall of Man is associated with freedom and the recognition of difference—that the man Adam differs from the woman Eve—and this individuation brings about time and death. More explicit than Plato in his *Symposium* is Plotinus in his *Enneads*: "What can it be that has brought the souls to forget the father, God, and, though members of the Divine and entirely of that world, to ignore at once themselves and It? The evil that has overtaken them has its source in self-will, in the entry into the sphere of process, and in the primal differentiation with the desire for self-ownership."[76]

Accepting "love" as a metaphor for entanglement indeed makes it possible to read the biblical Genesis as an allegory for quantum decoherence, but in medieval Francia, such thoughts were suspicious. "Irish porridge," judged the council at Valence in 855, condemning Eriugena's treatise.[77] After all, an all-encompassing unity as a concept for God threatens the clerical monopoly. If God is everywhere, no cast of priests is required as an intermediary for the believer to get in contact. Consequently, Eriugena's *Periphyseon* was included in the condemnations of 1050, 1059, 1210, and 1225, and after the first printing of the *Periphyseon* appeared in Oxford in 1681, it was almost immediately placed on the Index of Prohibited Books three years later.[78]

Nevertheless, as a protégé of the king, Eriugena remained untouched, even as his work was proscribed. The argument between Gottschalk, Eriugena, and the official church stands as a prime example of the contradictory traditions of Christianity, torn between dualism and Platonism, while at the same time claiming a political role. It is also a baffling document showing how prevalent monistic philosophy was over the course of history—which makes it even harder to understand why monism wasn't considered seriously as an interpretation of quantum mechanics for so long.

The Rationalist Made Mystic

Four hundred and fifty years later, the Christians' scorn of monism had intensified so much that it could turn against a high-ranking monk and theologian, to his own consternation. When Meister Eckhart was born in 1260 CE, in part due to the reforms of the Carolingian Renaissance, education, wealth, and commerce had revived in western Europe, and urban centers with an emerging middle class were flourishing. The first European university got established in Bologna in 1088, followed by Paris in 1150 and Oxford in 1167. New foundations in the thirteenth century included those in Cambridge and Padua, founded by renegade scholars from Oxford and Bologna in 1209 and 1222, respectively. In the fourteenth century there were already almost fifty universities in Europe, including in Pisa, Prague, Heidelberg, and Cologne. The previous decades had been influenced by the reign of Emperor Frederick II, the highly educated and religiously tolerant grandson of both the German emperor Frederick Barbarossa and the Norman king Roger II. Having grown up in Sicily, which his Norman ancestors had conquered from the Arabs in the eleventh century, Frederick had been versed in Arabic, Greek, and Latin since his childhood and harbored a keen interest in science and mathematics. After he had donated some books on Aristotle's natural philosophy to the university in Bologna in 1232, Albertus Magnus introduced them to the curriculum of the "seven liberal arts" that all students had to master before they became lawyers, physicians, or theologians. About two generations later, Eckhart studied, preached, and lectured in Cologne, Strasbourg, and Paris, while the Cologne Cathedral, the Strasbourg Cathedral, and Notre-Dame de Paris were built. During his lifetime Marco Polo set off to travel to eastern Asia, Jesus's alleged crown of thorns was transferred to Paris, and Eckhart's mentor, Theodoric of Freiberg, explained the formation and colors of the rainbow, relying on geometry and experiment.

Yet the new wealth and knowledge and the elevated confidence of western Europe also provoked tensions. These included new wars between the Christian and Muslim empires, conflicts between the rich and the poor, secular and clerical power, and Christians and Jews, and a clash between reason and faith, culminating in crusades and massacres. Already the First Crusade of 1096 had begun with the brutal Rhineland massacres of Jewish minorities, committed by a mob of thousands of predominantly poor Christian volunteers on their way through eastern France and western Germany. It had ended with the conquest of Jerusalem in 1099 and the slaughter of the city's civilian population. No better was the Fourth Crusade, which culminated in the sack and massacre of Christian Constantinople in 1204: sanctuaries were violated, tombs were plundered, works of immeasurable artistic value were stolen or destroyed, and thousands of civilians were raped or killed in cold blood. The sack weakened the Byzantine Empire permanently and contributed to the rise of Venice in the West and the Turkish Ottoman Empire in the East. When Frederick II hesitated to embark on a promised crusade, Pope Gregory IX excommunicated the emperor in 1227 and later even vilified him as the "Antichrist."

At the same time an increasing number of believers criticized the wealth and the mundane and depraved demeanor of the official Catholic Church. Mendicant Christian orders such as the Dominicans and Franciscans preached a lifestyle of poverty and care for the increasing numbers of homeless and sick. For similar reasons, starting in the early thirteenth century, the Cathars, a dualist sect in the Manichaean tradition, had gained importance in southern France. For the Cathars, everything earthly was the work of Satan, including matrimony, sex, property, animal products, and the official church. Gregory's predecessor, Innocent III, employed the Dominican and Franciscan monks in his effort to eradicate these heretics and for the first time declared a crusade against a region within the throes of Christian Europe, culminating in a genocide. "Kill them all, and God will know his own," the papal legate allegedly commanded when the crusaders took the city of Béziers and slaughtered nearly every man, woman, and child in the town.[79] After armies had devastated southern France for two decades and the Cathars still hadn't been entirely eradicated, Pope Gregory IX finally instituted the papal Inquisition in 1233, which sought out and persecuted dualists, monists, and scientists alike in the Catholic Church's entire sphere of influence for the next five hundred years. Gregory's chief inquisitor for Germany, Konrad of Marburg, alone made sure that hundreds of alleged heretics were burned alive at the stake. In this climate, blood libels spread that falsely accused Jews of ritual murder and provoked pogroms. Even black cats were almost eradicated throughout Europe since Gregory IX believed them to be the devil in disguise.

Meanwhile, rediscovered Greek and Arabic texts inspired a new generation of natural philosophers in Paris and Oxford. At the University of Paris, rifts were growing between the Aristotelian liberal arts and Augustinian theology. The teaching of Aristotle's physics was forbidden in 1210 by the bishop of Paris, in 1215 by the papal legate and chancellor of the University of Paris, and in 1231 by Pope Gregory IX himself, only to reappear on the mandatory reading list in 1255. In this situation a group of scholars in Paris had even resorted to the view that there exists a double truth: that what is right in natural philosophy may be wrong at the same time in theology and vice versa. This paradoxical doctrine constituted only an extreme example for the split between God and world that Christianity inherited from its origins as a counterreligion—a medieval precursor of the "Shut up and calculate" motto that became prominent in quantum mechanics. The reconciliation of these arguments by Albertus Magnus's student, the later saint Thomas Aquinas, finally resulted in the radical pigeonholing of the Scholasticism that strictly discouraged any statement about the fundamental reality. This division of the scopes of application later got lucidly illustrated in Raphael's fresco *School of Athens* in the Apostolic Palace in Rome. By depicting Aristotle, pointing down to Earth, next to Plato, who points up to the heavens, the painting anticipated Bohr's vertical complementarity between the projector and on-screen realities in quantum mechanics, or Heisenberg's reasoning about religion and science applying to different aspects of the world—beliefs that originated in the thirteenth century.

In 1277 the bishop of Paris condemned 219 specific theses. According to his decree, it wasn't permissible to argue for determinism, atomism, materialism, or the unreality of time or that only philosophers are wise men. Nor were Christians permitted to question authority, the afterlife, the reasonableness of prayer and confession, the sinfulness of homosexuality and extramarital sex, or the justification of killing animals. This multifaceted catalog was an extended and more detailed version of a list of ten heresies the bishop had published seven years earlier, which prominently reprobated the claim that "God doesn't know singulars"—that for God, in other words, all is One. Indeed, already at the dawn of the thirteenth century, Amalric of Bena, a lecturer at the University of Paris, had appropriated Eriugena's philosophy and taught that "all is one, and all that is is God."[80] In 1204 Amalric was condemned as a heretic. Six years later his remains were exhumed from his grave and cast into unconsecrated ground, while ten of his followers, who were reportedly rejecting the sacraments and practicing free love, were burned at the stake in Paris.[81] One of these convicts, it is said, drove his prosecutors up the wall by claiming he couldn't be burned by fire or tormented by torture since as long as he existed, he would be God. Curiously, this heresy seemed important enough that Thomas Aquinas felt compelled to stress that a statement such as "a stone is God" had to be rejected.[82] More than seven hundred years later, the Catholic Encyclopedia of 1913 still judged that the pantheism of the Amalricians "of itself would justify the drastic measures to which the Council of Paris had recourse"; that "the University of Paris was being made the scene of an organized attempt to foist the Arabian pantheistic interpretation of Greek philosophy on the schools of Latin Christendom"; and that "in view of these conditions . . . the complete extirpation of the sect of amalricians . . . cannot be judged untimely or intemperate."[83]

In contrast to the Amalricians, Eckhart was particularly anxious to emphasize that man is God only as long as he lives a just life—but to no avail. It is enough that Eckhart's reasoning was focused on interrelations and the whole and thus burst the tight bonds of Scholasticism to get him into trouble. Eckhart argued, for example, that justice isn't a property belonging to a just person; quite to the contrary, the just person is a participator in global justice and thus in God. Even worse, his treatises and sermons betrayed his monistic influence. To Eckhart, "God . . . is One in his hidden unity" and "flows into all things."[84] He is "in the innermost part of each and every thing."[85] Accordingly, "we should grasp God in all things," and "if we take a fly as it exists in God, then it is nobler in God than the highest angel is in itself," as Eckhart preached.[86]

Thus—although Eckhart was more cautious than the Amalricians, and although he was a high-ranking member of the Order of Dominicans, known for its learnedness and for serving as the backbone of the Inquisition—he nevertheless was condemned. When the archbishop of Cologne issued a first verdict against him, Eckhart denied the archbishop's authority and appealed to the pope. At almost seventy years of age, he set off on foot to walk the 550 miles to Avignon, where the popes resided after they came under French influence in the aftermath of their conflict with Frederick II. Eckhart died soon after his arrival, but his process continued, ending with a papal bull determining that Eckhart had been seduced by the devil "to want to know more than necessary."

Although Eckhart himself wasn't convicted, twenty-eight of his theses were identified as heretical, and the books containing them were forbidden. Eckhart is remembered as a mystic, although he advocated a rational approach to the universe. And for another one hundred years, monism remained a no-go in the Catholic Church.

The Diplomat of Monism

Nicholas of Cusa (1401–1464), or "Cusanus," as he is known in Latin, was born in the small town of Cues, on the bank of the Moselle, one of Germany's finest wine regions, as the son of a shipowner. It became clear early on, however, that it wasn't his calling to continue the family business. According to a local anecdote, his angry father threw him overboard once since he was constantly reading instead of rowing. Even so, his family was prosperous enough to send him to the University of Heidelberg when he was fifteen years old. This marked the beginning of a career that, in many aspects, made history.

The timing of his education was fortunate indeed. Just one year later the Tuscan Humanist, papal secretary, and avid book hunter Poggio Bracciolini rediscovered Lucretius's *On the Nature of Things* (*De rerum natura*), probably in the Benedictine library in Fulda, changing European culture forever.[87] Lucretius's book praises a naturalistic worldview that begins with a hymnic invocation of Venus, the Roman goddess of love and yet another incarnation of Isis, the monistic personification of nature.

While studying law at the University of Padua, Nicholas was brought up to date with the recent developments in science, read newly retrieved ancient Greek books, and got acquainted with some of the most important Italian families. His new friends included the mathematician Paolo dal Pozzo Toscanelli, who advised Filippo Brunelleschi when he designed the dome of the Florence Cathedral. After receiving his doctorate, Nicholas returned to Germany, working as a lawyer for the prince-archbishop of Trier and combing libraries for forgotten ancient books in his spare time.

On the occasion of an Epiphany celebration in 1431, Nicholas gave a remarkable speech that praised the biblical Magi as sages who may be found in all nations and compared them with Plato. This speech reflected two of Nicholas's lifelong, primary concerns: his efforts to achieve conciliation between religions and nations and his interest in monistic philosophy. When the Byzantine emperor requested military support from the pope in his defense against the Turks, Nicholas, whose brilliant speeches had already attracted attention, was appointed to sail to Constantinople to accompany the Byzantine emperor and his representatives to the papally summoned Council of Ferrara. For Nicholas, it must have been the trip of a lifetime. During the stormy cruise back to Venice, Nicholas shared his ship with three Greek philosophers who turned out to be kindred spirits: Gemistus Pletho and his students Bessarion and John Argyropoulos. Becoming friends with the Greek scholars in their exotic apparel, Nicholas must have felt as if the biblical Magi he had preached about seven years earlier had come alive. In his sermon, he had compared the Magi to Greek philosophers who were researching God's essence in nature. Now, while looking onto the storm-lashed sea, Nicholas experienced what he later described as a divine illumination inspiring him to think about complementary aspects of the One.

In his book *On Learned Ignorance* Nicholas describes "the one eternal thing," which he equates with God: "God is the one most simple Essence . . . of the whole world, or universe."[88] Like the quantum mechanical wave function, for Nicholas, God is "all that which can be."[89] Like Eriugena before and H. Dieter Zeh after him, he pondered "how it is that the oneness of things, or the universe, exists in plurality and, conversely, the plurality [of things] exists in oneness."[90] What is understood as entanglement and decoherence in the context of quantum mechanics, Nicholas described as the "enfolding and the unfolding of all things," emerging from a hidden reality that "while remaining incomprehensibly above all the senses and every mind . . . appear[s] . . . manifoldly in the different images multiplied from it."[91] Cusanus's One is the fundamental reality, the "oneness of the universe" that "is the root of all things."⁹² Following Plato's *Parmenides*, Cusanus also developed a notion of complementarity, when he argued that "the creation as creation cannot be called one, because it descends from Oneness, nor [can it be called] many, since its being derives from the One" and that "something similar, it seems, must be said about simplicity and composition and other opposites."⁹³ Finally, Cusanus's One isn't a bleak, transcendent concept; it embraces nature, since "anything whatsoever in the One," and thus the diversity of the world, "is understood to be the One."⁹⁴ As Cusanus explained, "Seeing the differences of things, we marvel that the one most simple Essence of all things is also the different essence of each thing."⁹⁵ Remarkably, he even credited the pagans with naming "God in various ways in relation to created things" and for also calling "Him Nature."⁹⁶ Indeed Nicholas believed that, in principle, pagans, Jews, and Christians all worshiped the same divinity: "The ancient pagans derided the Jews, who worshiped one infinite God of whom they were ignorant. Nevertheless, these pagans themselves worshiped Him in unfolded things."⁹⁷ Finally, Nicholas combined this monistic philosophy with a surprisingly modern conception of science, based on observation and a mathematical description of nature.⁹⁸

It is a consequence of both Nicholas's diplomatic skills and the times he lived in that these thoughts did not provoke any conflict with the church. On the contrary, Nicholas ascended to become a cardinal and later was even chosen as the pope's proxy in Rome. After their arrival in Italy—an event in the history of the Renaissance that may have been equally pivotal as the discovery of Lucretius's book twenty years earlier—Nicholas and his new friends Pletho and Bessarion moved on to attend the Council of Ferrara. The city was, however, threatened by the plague. On invitation by the rich Italian banker and politician Cosimo de' Medici, the meeting was relocated to Florence, where it changed the course of history.

Pletho provided Nicholas's friend Toscanelli with a first-century Greek map, which indicated a westward sea route to a land mass that was allegedly Asia. Toscanelli later sent it to Christopher Columbus, who carried it with him on his first voyage to America. As if this weren't enough, Pletho also lectured about Plato and inspired the enthusiastic Cosimo de' Medici to appoint the son of his physician, Marsilio Ficino, to found a new Platonic Academy to translate all of Plato's works and those of Plotinus into Latin. Later Cosimo appointed Ficino as tutor for his grandson Lorenzo de' Medici. This circle around Lorenzo "The Magnificent," Ficino, and his students included masterminds such as the painters Sandro Botticelli and Michelangelo, the poet Girolamo Benivieni, and possibly also the polymath Leonardo da Vinci.

Together these artists and scholars infused Renaissance thinking with monistic philosophy. Inspired by Plato, Plotinus, and philosophical texts known as "hermetica," Ficino and his student Pico della Mirandola developed a blend of Christianity with Egyptian and Greek monism that identified Plato's One with both nature personified by Isis or Venus and the Christian God. This notion is most beautifully realized in Botticelli's painting *The Birth of Venus*, which depicts the nude goddess of love and nature, blown over the ocean (an allegory of the One) to the shores where the Horae, minor deities of the seasons and time, await her with a cloak to cover and veil her when the goddess starts to live her corporeal and temporal life as unfolded nature. In the ensuing period, first Italy and then the rest of western Europe burst into an explosion of classically inspired architecture, art, and science—a revolution inspired by an appreciation of humanity and an embrace of nature, understood as the manifestation of an underlying unity or One. The new mind-set sparked "a doctrine that advocated harmony and tolerance," as the historian of philosophy Paul Oskar Kristeller writes, most evident maybe in the distinctly Platonic *Oration on the Dignity of Man* by Ficino's student and friend Pico della Mirandola.[99]

The world was becoming more modern, but the flowering of humanism, science, and the arts did not proceed in a straight line. The riches generated by the new sciences and economic developments of the Renaissance came to be distributed increasingly unfairly. Lorenzo the Magnificent's son Giovanni, the later Pope Leo X, became a living symbol of corruption, greed, and lavishness. As Paul Strathern describes, Giovanni enjoyed banquets with "dozens of courses, each served and eaten off different sets of silver and golden platters. Some of the dishes also incorporated trivial spectacles—nightingales flying from pies, young boys dressed as cherubs emerging from puddings . . . [H]is extravagances were legendary, and were indulged on an imperial Roman scale, perhaps best exemplified by his favourite dish, which was peacocks' tongues."[100] The excessive sale of indulgences, expedited by Leo X, to cover, among other things, the enormous cost of building St. Peter's Basilica, later drove Martin Luther to nail his theses to the door of the church in Wittenberg, finally leading to the Reformation. Giovanni's elder brother Piero became Lorenzo's successor as ruler of Florence and struggled with incompetence, hubris, and a lack of interest, until he was eventually driven out of Florence by the fundamentalist preacher Girolamo Savonarola. Piero never returned to his hometown, although he collaborated with the ruthless Cesare Borgia to regain power and poisoned Pico della Mirandola and his lover, the philosopher Angelo Poliziano, who had sympathized with Savonarola. Meanwhile Savonarola's zealots stripped the citizens of Florence of their possessions, including cosmetics, jewelry, art, musical instruments, and books, only to burn these valuable items in so-called bonfires of the vanities. Savonarola, after finally being executed by Florentine citizens, became a martyr of the Protestant Church.

Monism had stirred for a moment, but dualistic ideologies became popular again, among both Catholics and Protestants. They included Luther's repelling antisemitism, the fanatic iconoclasm of the Calvinists, and the increasingly brutal actions of the Inquisition and gave way to religious war, a new wave of terror against heretics and unbelievers, the widespread murder of alleged witches, and the antiscientific backlash of the Counter-Reformation.

The Martyr

At dawn on February 17, 1600, the troubled relationship between monism and the Catholic Church reached its gruesome climax. After having been offered a last breakfast of almond biscuits dipped in marsala wine, Giordano Bruno was silenced with a leather gag. Evidently, after he had spent eight years in the prisons of the Inquisition, the convict's sharp-tongued talk was still frightening his prosecutors. Next he was mounted on a mule that carried him to the pile of wood stacked on the Campo de' Fiori, the Roman marketplace and execution ground. Upon arrival, Bruno was stripped naked, tied to the stake, and burned alive while a group of black-hooded friars prayed, sang litanies, and begged him to recant. His last deliberate move was to turn his head away from the crucifix held before his eyes when he mounted the stake. They dumped his ashes into the Tiber.

Debates continue to this day about whether Bruno was a martyr for science, philosophy, or merely free speech. In his works, Bruno championed Nicolaus Copernicus's heliocentric model of the planetary system and painted the picture of an infinite universe permeated by divine spirit—an idea he adopted from Cusanus. As has been pointed out, for example, by Alberto Martínez, while Bruno wasn't a scientist in the sense that he performed experiments, observations, mathematical analysis, or model building, his conceptions about the universe were more accurate than those of Copernicus, Galileo Galilei, or Johannes Kepler. Bruno drew from many disciplines, but like so many other figures in these pages, he saw them as all one.

Just like Dionysius, Eriugena, Meister Eckhart, and Nicholas of Cusa before him, Bruno had based his philosophy on a monistic foundation. In his book *Cause, Principle and Unity*, Bruno enthusiastically affirmed "the thesis of Heraclitus, which declares all things to be but one" that "contains all things in itself."[101] Glowingly, Bruno described "the supreme perfection, the supreme beatitude [that] consists in the unity which embraces the whole," as "the colour which embraces all colours," the "sound, not . . . any particular one, but . . . [the] complex sound which results from the harmony of many sounds," "the one which is itself the all."[102] In other words, "the whole is one, as Parmenides . . . perhaps conceived it," and "there is no . . . part that differs from the whole."[103] Again, Bruno's monism is accompanied by a high esteem for nature: "Nature ... is none other than God in things," he wrote in *The Expulsion of the Triumphant Beast*.[104] From Plato Bruno adopted the image of the forest as an allegory for the material world, "where divine footprints lie hidden but when we take notice of them by means of reason ... the divine light" becomes accessible.[105]

Clearly, Bruno is embracing a pagan cosmo- or pantheism here. Sure enough, he credits Egypt as "the image of heaven"[106] and invokes Isis to explain "how a simple divinity that is found in all things, a fertile nature, preserving mother of the universe, shines forth in different subjects ... and takes on different names."[107] Bruno even went so far as to assert that David of Dinant, a heretic in the tradition of Almaric of Bena, "was not led astray by taking matter to be an absolutely excellent and divine thing."[108] Far from despising earthly matter, Bruno believed "the essence of the universe is one both in the infinite and in anything taken as a member of the universe," "the whole and each of its parts are but one," and "Parmenides was ... right to say that the universe is one."[109] Obviously, Bruno here undermined the separation of world and God that is so fundamental for monotheism in general and Christianity in particular. But though both Bruno and Nicholas of Cusa had been advocating strikingly similar thoughts, Nicholas flourished while Bruno burned.

In contrast to Nicholas, Bruno definitely wasn't known for his diplomatic skills. Already as a novice in the Dominican monastery in Naples, he had begun to question some of the doctrines of his Catholic faith and was consequently reported to the Inquisition for the first time. When, ten years later, an Inquisitor found a forbidden book with Bruno's notes in the margins in the monastery's latrine, Bruno decided to flee. He embarked on an odyssey in search of a place to work and to live in liberty and security, a voyage carried out on foot and astride donkeys through a major part of Europe as it was increasingly riven by religious conflicts—incidents such as the St. Bartholomew's Day Massacre in Paris in 1572, when thousands of Protestants were murdered by Swiss mercenaries, and Catholic mobs were signaling the coming of the Thirty Years' War that would kill almost half of the population of central Europe.

On these travels, Bruno's belligerence regularly put him in conflict with local intellectuals whom he liked to address as "pedant asses."[110] When passing through Geneva he criticized a professor, was refused the right to take sacrament, and was thrown into prison until he apologized on his knees. He was luckier in France, where he was asked to demonstrate his art of mnemonics at the court of King Henry III. After moving on to England, he was ridiculed and accused of plagiarism in Oxford. In Frankfurt, the city of the great book fair, he published poems and anticipated the controversial insight of quantum cosmology that time is an illusion: "Past time or present, whichever you happen to choose, or the future: All are a single present, before God an unending oneness."[111] Finally, Bruno returned to Italy where he lectured briefly in Padua, hopeful that he might be offered a professorship, but Galileo Galilei, six years his junior, received it instead. After that, Bruno took the fatal decision to accept an invitation to Venice, where his host would denounce him to the Roman Inquisition.

The Inquisition eventually justified Bruno's death sentence with his refusal to renounce eight propositions. These included the questioning of Catholic doctrines—such as that bread transmuted into flesh during Mass, that Christ performed real miracles, or that the persons in the Holy Trinity were distinct—as well as outright challenges to the church itself, such as the desire to create his own sect, where he could practice the art of divination as well as the use of Pythagorean motifs such as that there exist many worlds in the universe. While none of these accusations was explicitly monistic, as Alberto Martínez points out in *Burned Alive*, "the Pythagorean thread is remarkable because it explicitly shows up in the works of many important figures in the Copernican Revolution, including Copernicus, Bruno, Kepler and Galileo."[112] Pythagoreanism and Platonism had been so strongly entwined since antiquity that they got increasingly hard to discriminate. In essence, both philosophies were fundamentally monistic.

There are other indications that Bruno's monism was a critical factor contributing to his sentence. Bruno was neither the first nor the last monist to end up on the stake. Four hundred years earlier, the Amalricians had been burned to death in Paris, and nineteen years later Lucilio Vanini, a philosopher who had anticipated biological evolution, advocated that apes and humans shared common ancestors, and preached a religion of nature, was accused of atheism. When his interrogators inquired whether he believed in the existence of God, Vanini allegedly plucked out a blade of grass and said, "Already this leaf proves the existence of God."[113] He was sentenced to have his tongue cut out, to be strangled at the stake, and to have his body burned. Almost exactly two hundred years later, Walt Whitman was born and would go on to write, "I believe a leaf of grass is no less than the journey-work of the stars, And the pismire is equally perfect, and a grain of sand, and the egg of the wren, ... And a mouse is miracle enough to stagger sextillions of infidels."[114]

It is also instructive to compare Bruno's sentence with Galileo's run-in with the Inquisition. Sixteen years after Bruno's death, Copernicus's book was suspended by the Inquisition. For more than two hundred years, it was only allowed to be published in editions that stressed it presented just a mathematical model but no statement about reality. In the same year, 1616, the very same Cardinal Bellarmine, who had already interrogated Bruno, warned Galileo to teach the heliocentric model not as truth but only as a hypothesis. Another sixteen years later, Galilei was finally officially accused of heresy and, after he recanted, sentenced to house arrest. These events indicate that the Inquisition was strongly concerned with prohibiting science from making statements about the foundational reality and with upholding the separation of world and God that monism threatened to bring down. H. Dieter Zeh later compared Galilei's persecution by the Inquisition with the opposition his and Everett's realistic interpretation of quantum mechanics faced from the physics establishment: "Galilei was prosecuted since he understood the Copernican worldview as real; and not only as a tool to perform calculations. Similar efforts to downgrade scientific insights are common today not only among creationists but also among many philosophers . . . and even most physicists."[115]

To this day Giordano Bruno isn't fully rehabilitated by the Catholic Church. His writings remained on the Index of Prohibited Books until 1966, when the index was formally abolished. In 2000, Pope John II regretted the use of violence in Bruno's case but maintained that Bruno's teaching was incompatible with Catholic faith. This attitude is reflected in the judgment of the Catholic Encyclopedia of 1913, which says of Bruno, "From the neo-Platonists he derived the tendency of his thought towards monism. From the pre-Socratic philosophers he borrowed the materialistic interpretation of the One. From the Copernican doctrine, which was attracting so much attention in the century in which he lived, he learned to identify the material One with the visible, infinite, heliocentric universe."[116] Based on Bruno's monism, the author eventually arrives at the harsh verdict that Bruno "failed to feel any of the vital significance of Christianity as a religious system" and that he was a man without "a trace of religion."[117]

* * *

The history of monism, together with its many implications for science and religion, as well as for arts and politics, is fascinating on its own terms, but naturally we can't do more than barely scratch the surface here. I hope I've made clear that neither Bohr's refusal to accept the foundations of quantum mechanics as "real" nor the general hesitation to accept the monistic implications of quantum mechanics can be traced back exclusively to a prevalent positivism among scientists of the early twentieth century, a vaguely substantiated reference to Bohr being under the spell of "German idealism," or an alleged widespread leaning against "causality, individuality, and . . . visualizability" prevalent among intellectuals in postwar Europe of the 1920s and in particular the Weimar Republic in Germany, as historian of science Paul Forman has argued.[118] For millennia, the official church had taken pains to prohibit monistic conceptions of nature and to push monism into an exclusively religious, otherworldly realm. Starting from the anonymous philosopher known as Dionysius, who became a church father, through Eriugena, whose books got forbidden, and Meister Eckhart, who was convicted of being possessed by the devil, to Cusanus, who climbed the highest ranks in the papal administration, and back to Bruno, who got burned at the stake, the church had struggled to appropriate monism in its concept of God, on the one hand, and to persecute vigorously, on the other hand, any conflation of God or monism with the natural world.

In the early twentieth century Bohr and Heisenberg surely didn't have to fear being burned at the stake or scorned by the church; nor were they particularly religious themselves. Yet somehow they, and many other scientists too, had internalized the narrative the church wanted to convey—namely, that monism and nature or monism and science don't belong together; that the hypothesis that "all is One" simply isn't proper science. It is one of the main concerns of this book to refute this allegation and to reclaim monism for science. Yet, as fascinating as it is that quantum mechanics first rediscovered a three-thousand-year-old philosophy and then promptly denied it, has this philosophy ever exerted any direct influence on science? And even if it did, could this age-old idea possibly be of any use in the challenges we are facing in fundamental physics today?

5

FROM ONE TO SCIENCE AND BEAUTY

WHAT IS SCIENCE, AND WHICH THEORIES AND HYPOtheses actually qualify as scientific? These questions are fiercely debated among scientists today, yet date back to the Renaissance and Enlightenment eras. In fact, much of our modern worldview originated in the great scientific revolutions since the sixteenth century that are commonly associated with the names of Nicolaus Copernicus, Galileo Galilei, Isaac Newton, Michael Faraday, and James Clerk Maxwell. Rarely mentioned in this context are monism and the works of Plato, Marsilio Ficino, Leonardo da Vinci, or Johann Wolfgang von Goethe. Even so, as an inspiration to search for unification and beauty in nature, monism became a powerful trigger for and catalyst of scientific creativity. It was (and still is) the monistic tradition inherent in Platonic and Pythagorean philosophy that drove scientific revolutions, much more so than the Aristotelean pigeonholing that characterized medieval thought.

From Platonic Love to Love of Science

During the late Middle Ages, the monism-infused books of antiquity, having survived the fall of Rome by being preserved in Constantinople and the Islamic world, slowly returned to western Europe. This trend had intensified in the fourteenth and fifteenth centuries, through trade relations with the flourishing Italian city-states and especially when Constantinople got conquered by the Turks in 1453. While the city was plundered for three days, as civilians were murdered, raped, and enslaved and churches desecrated and destroyed, a small fleet of ships escaped and brought refugees and precious scrolls to Italy. Books, ideas, and influences were sweeping western Europe, to be copied over and over again by the printing press invented around the same time by Johannes Gutenberg. These stimuli not only brought back an awareness of Greek philosophy but demonstrated the imperfections of the medieval tradition and inspired a new critical spirit. As a consequence, they helped spark the Renaissance age and the subsequent explosion of the arts and sciences. And they tied the philosophy of "the One" back again to science.

Importantly, in contrast to some trends in Platonism, and in particular medieval Christianity, the monism of the Renaissance didn't look down on nature. Instead, it embraced nature as a manifestation of the inaccessible "One." Spreading out from Florence and Marsilio Ficino's Platonic Academy, monism began to inspire arts and poetry around the globe. In addition, Ficino's circle evolved into a role model for later scientific academies, such as the Accademia dei Lincei in Rome (founded in 1603), which boasted Galileo Galilei as a member, the Accademia del Cimento in Florence (founded by Galilei's students in

1657), and eventually the Royal Society in London (1662) and the Académie Royale des Sciences in Paris (1666).[1] As the accomplished Renaissance scholar Paul Oskar Kristeller has pointed out, "If we wish to understand the notable difference which separates the philosophical and scientific thought of Bacon, Galileo and Descartes from that of St. Thomas Aquinas or of Duns Scotus . . . Florentine Platonism will occupy an important place."[2] In subsequent centuries and inspired by monistic vibe, Galileo Galilei realized that all objects fall with the same speed, Isaac Newton discovered that the same physical laws govern both motion on Earth and the passage of planets and stars across the sky, and Baruch Spinoza equated "God" and "nature" as a single, eternal, and necessarily existing substance.

A central element of Ficino's Florentine Platonism was the concept of Platonic love. The idea goes back once again to Plato's *Symposium*, which actually doesn't end with Aristophanes's mythological narrative about lovers seeking in each other their lost halves. Aristophanes's fable only sets the stage for Plato's final speaker, Socrates, who elevates the unification in the act of loving to a universal philosophy of life that became known as "Platonic love" in the Renaissance. Often misunderstood as a sterile, asexual friendship, Platonic love in fact describes a way of loving where love for a specific individual only reflects a deeper love for the entire universe.

It is a message with explicit scientific ramifications. Plato explicitly addresses the researcher who "will go on to the sciences, that he may see their beauty, being not like a servant in love with the beauty of one youth or man or institution . . . , but drawing towards and contemplating the vast sea of beauty."[3] Through science, the scientists "will suddenly perceive a nature of wondrous beauty," a "beauty absolute, separate, simple, and everlasting . . . without . . . any change" that is "imparted to the ever-growing and perishing beauties of all other things."[4] Following Plato's *Symposium*, "Ficino treats love as a cosmological principle of the unity of things," Kristeller explains, and determines that this monistic philosophy developed into a crucial momentum of sixteenth-century philosophy.[5]

Ficino's monism—while in its original form still mashed up with astrology, alchemy, and other forms of superstition—eventually fueled the scientific revolutions that made the world modern. At least three important threads can be identified in which the mind-set to take the part as representative of the whole, so characteristic of monistic philosophies, begets the paradigm, practices, and narratives that signify modern science until the present day.

First, Renaissance monism implied a new appreciation of nature that eventually manifested itself in empiricism, the foundation of modern science that stresses the importance of sensory experience in general and controlled experiments in particular. Increasingly, nature was seen not as just a subordinate realm for humanity to exploit and have dominion over but as worthy of study for its own sake.

Next, it was expected that observable nature would reflect the harmony of the unifying One. Given that the Renaissance monists still were devout Christians, this esteem for nature encouraged a trend toward "natural theology," the idea that God could be recognized in his creation. This view was often illustrated by the notion that there was a "Book of Nature" that offered an equally good or even better glimpse of the divine than scripture. A growing number of Renaissance and Enlightenment philosophers and scientists now understood themselves as deists, who thought of God as a creator who could be understood only rationally and revealed himself exclusively in nature instead of in miracles or revelations, if not as pantheists, who equated God entirely with the universe. Obviously, such a theology was perfectly suited to revive ancient Platonic and Pythagorean doctrines to search for symmetries and harmony in the cosmos. Even if not all ideas inspired by this vision turned out to be fruitful or correct, this search for simple, rational, and aesthetically pleasing explanations of natural phenomena evolved into a lasting influence on how science is practiced until today.

Finally, the monistic underpinning supported a trend to search for universal laws for the most different realms of nature. After all, it isn't obvious a priori that the same laws of nature exist in different regions and different epochs of the universe, and in fact such a unifying philosophy contradicted the medieval tradition. As Kristeller writes, "The medieval conception of the universe was dominated by the idea of a hierarchy of substance." In contrast, "in the astronomy of Kepler and Galileo there is no room for differences of rank and perfection between heaven and earth, or between the various stars or the various elements."[6] Kristeller attributes the "gradual disintegration of the old idea of hierarchy" to the monistic philosophy of Nicholas of Cusa, Giordano Bruno, and Ficino that sought "a central link which . . . could mediate between the opposite extremes of the universe."[7] The idea of unification (i.e., that a minimal set of natural laws governs everything that happens, has happened, and will happen in the universe) is an obvious heritage of conceiving nature as "One."

These considerations offered more than a charming narrative frame. There were indeed monistic elements inherent in the genuine methods contributing to scientific progress up to the point where they were directly triggering scientific revolutions.

A Disciple of Nature

In the course of the Renaissance era, this blend of monistic philosophy with various Neoplatonist and Pythagorean concepts evolved into a powerful motif engendering the Scientific Revolution and modern science ever since. One prominent exemplar was the Tuscan polymath Leonardo da Vinci, who represented the Renaissance spirit like no one else.

Da Vinci considered himself a "disciple of nature."[8] He was a vocal advocate of exact observation: "Sound experience" was to him "the common mother of all the sciences and arts."[9] Yet this commitment, as revealed in his meticulous anatomic studies and his precise drawings of plants, geometrical objects, and astronomical phenomena, wasn't at odds with his vision of nature being governed by reason and harmony: "Although nature herself begins from the reason and ends in the result, we must pursue the contrary course and begin . . . from experience and by it seek out the reason," da Vinci wrote.[10] According to his biographer Walter Isaacson, Leonardo "had a reverence for the wholeness of nature and a feel for the harmony of its patterns, which he saw replicated in phenomena large and small."[11] In his notebooks, Leonardo enthusiastically praised "the beauty of the universe," "the things in nature" that he felt "beget a harmonious concord in one glance,"[12] a beauty governed by the laws of mathematics: "Let no man who is not a Mathematician read the elements of my work,"[13] Leonardo insisted (adopting a Platonic motto), emphasizing, "There is no human experience that can be termed true science unless it can be mathematically demonstrated."[14] As his famous drawing *The Vitruvian Man* demonstrates, da Vinci was fascinated by the relations between geometry and natural proportions. For example, Leonardo explained the beauty of a face with "the divine proportion of the limbs united one with another, . . . [that] compose . . . [a] divine harmony."[15] This fascination also gets reflected in Leonardo's illustrations in his friend, mathematician Luca Pacioli's book, which identified the divine proportion with a specific number, the "golden ratio." The golden ratio is defined by the observation that there is only one possible way to divide a line into two parts such that the ratio between the length of the longer part and the length of the shorter part equals the ratio between the sum of the two parts' and the longer part's lengths. The number that pops out for the ratio is irrational (it takes infinitely many digits to write it down) and unique, and it occurs copiously in mathematical series, in geometrical objects such as cubes and polyhedrons, and in works of arts and nature.

Like Ficino and Giordano Bruno, da Vinci imagined Earth or even the universe as a living creature, an organism. For Leonardo, the human body was "a lesser world," a miniature embodiment of the entire universe: "As man has within himself bones as a stay and framework for the flesh, so the world has the rocks that are the supports of the earth. As man has within him a pool of blood wherein the lungs as he breathes expand and contract, so the body of the earth has its ocean."[16] To him "every whole is greater than the part," while "the whole [still is present] in every smallest part": "each in all and all in every part," da Vinci summarized this philosophy.[17] Alexander von Humboldt later credited Leonardo with having been "the first to start on the road towards the point where all the impressions of our senses converge in the idea of the Unity of Nature."[18]

Da Vinci's monism may be traced back at least partly to Florentine Platonism. He had been born near Florence and got educated in the workshop of a Florentine painter, where he must have been in contact with Sandro Botticelli and Ficino. Later, he continued his career in Milan and Rome and eventually moved to France, upon the invitation of King Francis I. While the monistic seed sowed with the Florentine Renaissance often brought its early advocates a premature death at the stake, it eventually led to modernity. Spread by merchants and other travelers, but most importantly by students who attended the Italian universities and then returned to their home countries, monistic philosophy and Platonic love became part of Europe's mainstream culture.

The Mother of All Scientific Revolutions

One of these students was a twenty-three-year-old, freshly ordained canon of the Frombork Cathedral in Poland. He was Nicolaus Copernicus—the man who would overthrow the traditional cosmology of Ptolemy and the Middle Ages. When Copernicus died in 1543, he held in his hands (so it is said) his book, *On the Revolutions of the Heavenly Spheres*, which would end up dislodging mankind from the center of the universe.

Copernicus's background, like da Vinci's, can be traced back to a monistic tradition. As a student in Bologna, Copernicus worked with the astronomer Domenico Maria da Novara, with whom he observed a lunar occultation of the star Aldebaran, which Copernicus later used to disprove Ptolemy's geocentric model. Novara had studied in Florence under Luca Pacioli, Leonardo da Vinci's friend who was so enamored with the golden ratio that he went as far as to relate the number to the uniqueness, incomprehensibility, and omnipresence of God himself. Novara also considered himself a student of the German mathematician Regiomontanus, who had spent four years living in the household of Bessarion, the Platonist philosopher who had been sailing from Constantinople to Italy together with Nicholas of Cusa and Pletho. Bessarion had later been appointed a cardinal and developed into a leading intellectual who assembled the largest private library of his time, which would become the heart of the famous Library of St. Mark in Venice. He also made an important contribution to astronomy when he asked Regiomontanus to produce a new, abridged version of Ptolemy's compendium of Greek astronomy, the *Almagest*, the book that was later used by both Copernicus and Galilei and remained an influential champion of Platonism.

Thus, in Italy Copernicus became up to date with the new astronomy that would lead him to reckon with the inconsistencies and problems of the geocentric model. According to André Goddu, the "direct contact with the Platonist revival" was "the second major development in Italy" experienced by the young man from Poland.[19] Copernicus began to read Platonic and Pythagorean authors, such as Ficino's translations of Plato's *Parmenides* and *Timaeus*, Plutarch, and Bessarion, and developed an "interest and admiration for Plato" that was "consistent with Bessarion's interpretations and with the Neoplatonic tradition in general," as Goddu points out.[20]

After continuing his studies in Rome and Padua, Copernicus finally returned to Poland, where his doubts about the traditional model of the planetary system, the scattered remarks about ancient authors who questioned the geocentric model, and the inspiration drawn from Platonic and Pythagorean notions must have coalesced into the conviction that the sun, rather than Earth, was at the center of the planetary system. In fact, this arrangement nicely matched the conception of some Platonists that the sun could be identified as a material image of the One. As Goddu points out, in an entirely Platonic or Pythagorean fashion, Copernicus saw his magnum opus, *On the Revolutions of the*

Heavenly Spheres, as "concerned with the most beautiful objects, most deserving to be known."[21] Yet, for more than twenty years, he didn't publish his work. As he later explained in his dedication to Pope Paul III, he "was in great difficulty as to whether [he] should bring to light [his] commentaries," given "the scorn which [he] had to fear on account of the newness and absurdity of [his] opinion."[22]

According to legend, Copernicus received his printed book on the very day that he died. Its title page carried the alleged motto of Plato's Academy that had already fascinated da Vinci: "Without knowledge of geometry, do not enter here." Its first part finishes with a foray into Pythagorean philosophy and secrecy. Indeed, the view that Earth moves would become generally known as a "Pythagorean doctrine," both by its advocates and its adversaries.[23] Johannes Kepler, for example, later called Pythagoras the "grandfather of all Copernicans."[24] And according to Alberto A. Martínez, it was exactly this tradition that brought Bruno and Galilei into trouble with the Inquisition: "The Inquisition condemned the Copernicans as a heretical 'sect' of New Pythagoreans."[25]

While Copernicus's model didn't allow for a better fit with observational data yet, as it missed the concept of elliptic orbits later found by Kepler, it nevertheless represented an important breakthrough by dispensing with the traditional hierarchical cosmology that assigned to Earth a distinguished place in the universe. After Copernicus, Earth could be understood as but one compound of a larger universe governed by the same laws of physics, a trend that continued in the science of Galilei and Newton and in the modern physics of the nineteenth and twentieth centuries. Monistic philosophy, it seems, helped Copernicus to abandon the anthropocentric view that humanity occupies the center of the universe. It may also have tempted him to constrain his cosmology to perfect circles and spheres, making it difficult to reconcile with observational data. As we will see, this blend of accurate observation or experimentation with a quest for mathematical beauty and unification would evolve into a trademark of science.

The Music of the Spheres and the Book of Nature

Copernicus's champion in Italy became Galileo Galilei. Galileo was born in Pisa to Florentine parents. His father, Vincenzo Galilei, was a well-known musician who was part of a group known as the Florentine Camerata, which strove to revive the music of antiquity, an endeavor that eventually led to the development of the opera and the beginning of the baroque period in music. A major source of inspiration and at the same time a matter of study in this activity was the Platonic and Pythagorean tradition: as Paul Oskar Kristeller points out, "The theorists of the Camerata quote Plato as an authority and were . . . influenced by his doctrines."[26] An idea that featured prominently in this philosophy was the hypothesis that nature, math, and music had an intimate relation. Again, this Pythagorean motif can be traced back to Marsilio Ficino, who was an enthusiastic lutenist and even compared himself to the ancient mystical singer Orpheus.[27] To Ficino, music fulfilled a universal purpose. As Kristeller explains, Ficino believed that "the human soul acquires through the ears a memory of the divine music which is found first in the eternal mind of God, and second in the order and movements of the heavens."[28] Vincenzo Galilei's teacher, the Venetian music theorist and composer Gioseffo Zarlino, is known to have adopted Ficino's views.[29] During an argument with Zarlino about the proper tuning system for musical instruments, Vincenzo Galilei set out to probe and eventually debunk the Pythagorean myth told, for example, in Boethius's *De musica* that string tension and tone pitch would be connected by the same linear relationship as string length and tone pitch. By resorting to "experiment, the teacher of all things," Vincenzo Galilei finally found that the correct relationship is quadratic, probably one of the oldest known nonlinear relations in physics.[30] Of course Vincenzo attributed this blunder not to Pythagoras himself but rather to his overzealous followers.[31] It is quite possible that Vincenzo's son Galileo was inspired by his father's work when he started to study nature experimentally. What is sure is that Galileo later continued and improved on his father's experiments on tone pitches in various instruments. And in fact, one of Galileo's most important achievements is the mathematical description of free fall and projectile motion, in which the vertical distance covered depends quadratically on time.

Besides being famous for his carefully planned experiments, Galileo Galilei is popular for his advocacy of mathematics and his references to the "Book of Nature." As he wrote in his *Assayer*, "Philosophy is written in this grand book, the universe, which stands continually open to our gaze. But the book cannot be understood unless one first learns to comprehend the language and read the letters in which it is composed. It is written in the language of mathematics, . . . without which it is humanly impossible to understand a single word of it."[32] Galileo's quest for unification is obvious, for example, from his discovery that in vacuo all objects, independent of their mass, shape, or composition, fall at the same speed. Likewise, Galilei discovered that Earth and the heavens weren't fundamentally different: when he pointed his telescope to the moon and found mountains and valleys, just like on Earth, he in fact revived "the old opinion of the Pythagoreans that the Moon is, as it were, another Earth," as he wrote in his *Sidereal Messenger* in 1610.[33]

Meanwhile, in Prague, at the court of Holy Roman Emperor Rudolf II, Johannes Kepler was desperate to catch a glimpse of Galilei's treatise. Ever since six years earlier, when he had observed the magnificent supernova appear that would later be named after him, he had been convinced that Aristotle's doctrine of the immutability of the heavens was wrong. When Galilei invited him to comment on his results, Kepler enthusiastically congratulated him for "having pierced the heavens."[34] But Kepler also sent critical remarks. Knowing that a unified cosmology had been advocated by monistic philosophers before, Kepler scolded Galilei for not giving credit to his predecessors, in particular Giordano Bruno,[35] who already in 1584 had anticipated that the moon had mountains and valleys.[36] In view of the fact that Bruno had been burned by the Inquisition only ten years before and that Galilei himself was soon to become a target of the Inquisition's investigations, he probably knew better. But Kepler also was much more committed to a monistic worldview than Galileo was. While Kepler was educated in Tübingen, Germany, to become a Protestant priest, he had started to read monistic philosophers such as Nicholas of Cusa and Giordano Bruno. Kepler never got ordained as a priest though; before he could finish his studies, he was sent by his university to a school in Graz, Austria, to teach mathematics. Writing on the blackboard while giving an astronomy class in Graz, Kepler had a revelation that changed the course of his life: it seemed that an equilateral triangle connected the conjunctions of Jupiter and Saturn (i.e., the locations where Jupiter and Saturn as viewed from Earth seem to pass each other). "Unlike the astronomers before him, who satisfied themselves with simply recording the observed positions of the planets, Kepler was seeking a theory that would explain it all," the Israeli American astrophysicist Mario Livio wrote of Kepler's motivation.[37] In a decidedly Platonic attitude, Kepler believed that "in all acquisition of knowledge it happens that, starting out from those things which impinge on the senses, we are carried by the operation of the mind to higher things which cannot be grasped by any sharpness of the senses."[38] Kepler immediately went on to check whether other regular polygons, such as the square, would fit the conjunctions of other pairs of planets. But since there exists an infinite series of regular polygons, this approach wasn't in any sense predictive.

Next, Kepler fit the planetary orbits with polyhedrons, the three-dimensional generalizations of polygons of which there exist only five regular versions, known as "Platonic solids." This looked more promising, but to really check his model against observations, he needed the best astronomical data available. These were the data of the belligerent Danish astronomer Tycho Brahe, an arrogant genius who had lost a large part of his nose in a sword duel with his cousin, after the two had quarreled about who was the better mathematician. Now, after a dispute with the Danish king, Brahe had abandoned his manor and observatory, Uraniborg, which had been constructed according to Pythagorean musical ratios on the Danish island Hven, and accepted the position of imperial astronomer to the Bohemian king and Holy

Roman Emperor Rudolph II in Prague. Meanwhile, Kepler had realized that just as there are five regular polygons, there are five harmonic intervals in music, and he tried now to employ musical intervals to describe the relations of planetary velocities. Kepler realized that the ratio of velocities was close to 1:2 for Jupiter to Mars, 3:4 for Saturn to Jupiter, 4:5 for Mars to Earth, 5:6 for Earth to Venus, and again 3:4 for Venus to Mercury.[39] In his charming scheme, which became known as "the music of the spheres," these frequency ratios translated into an octave, a fourth, a third, a minor third (so that the velocities of Mars and Earth produce a fifth), and another fourth. Taken together, the six planets occupying the closest orbits to the sun conspired to produce a major chord spanning two and a half octaves. In the coming winter, Kepler, eager to test his theory against data, left Graz to join Brahe in Prague, though it wasn't until after the latter's death that he could access the secretive astronomer's notes.

In the following years, Kepler pursued his Pythagorean vision of mathematical beauty through personal hardships and troubled times. In 1611, Kepler lost a beloved son and his first wife. In 1612 Emperor Rudolf II, who had been forced to resign the year before, died, and Kepler had to leave Prague. In 1613, Kepler married again, but the first three children of this marriage died in childhood. Between 1615 and 1621 Kepler's mother was accused of witchcraft, incarcerated, and chained up for fourteen months. Even though Kepler finally obtained an acquittal, his mother died half a year after she was released. In 1618, the Thirty Years' War broke out, after Kepler himself had been threatened many times by religious tension and upheaval, from both Protestants and Catholics. Yet, in 1619 he finally published his *Harmonies of the World*, the book that contained his famous laws of planetary motion and replaced Copernicus's circular orbits with elliptical trajectories to explain the planetary velocities.

For a long time, after Newton's mechanics allowed for arbitrary elliptic orbits depending only on the initial conditions and the central star's and satellite's masses, Kepler's original vision about "the music of the spheres" was seen as at an impasse, "not only absolutely wrong, but... crazy even for Kepler's times," as Mario Livio states,[40] or as "the odd and unlikely midwives to Kepler's 'new astronomy,'" as Kitty Ferguson writes.[41] Yet systems involving three or more bodies, such as the solar system with its eight planets, can exhibit chaotic behavior where small distortions can have catastrophic consequences. The famous "butterfly effect," positing that the flap of a butterfly's wings may cause (or prevent) the formation of a tornado, illustrates this behavior. In this context, relations of orbital velocities corresponding to ratios of small natural numbers (such as Kepler's celestial music) or very irrational numbers (such as the golden ratio) characterize resonant or nonresonant behavior that determines whether a specific orbit is stable over long time spans or whether the planet sooner or later will be ejected from the solar system. Thus "Kepler's intuition wasn't so wrong after all," physicists Peter Richter and Hans-Joachim Scholz concluded in a 1987 essay discussing occurrences of the golden ratio in nature.[42] What's more, manifestations of universal harmony have been discovered elsewhere: a more faithful version of celestial music with literally cosmic dimensions can be found in the fluctuations of the early universe's primordial plasma. As cosmologists Wayne Hu and Martin White expounded in *Scientific American*, acoustic oscillations of the various cosmic components such as ordinary and dark matter that got imprinted (and convey important cosmological information) in the cosmic microwave background can best be understood as a harmony of tones and overtones constituting a real "cosmic symphony."[43]

God aka Nature

After the trials against Bruno and Galilei, scientists came under increasing scrutiny in Italy and other Catholic countries. As a consequence, the hot spots of science moved to northern Europe. Already the telescope that Galilei used to "pierce the heavens" had been invented before by a Dutch spectacle maker. Later, Galilei's final book and scientific testament, *Discourses and Mathematical Demonstrations Relating to Two New Sciences*, was published in Amsterdam, since it was forbidden in Catholic countries. When Galileo's students established a scientific academy in Florence, it lasted only ten years, until its patron, the Tuscan prince Leopold, became a cardinal. The shutdown of the academy apparently was a condition of his appointment.

Meanwhile the Netherlands, located strategically at the mouth of the Rhine and thus controlling the trade with the German backcountry, as well as a part of the sea route from the Baltic to the Mediterranean, were rising as a world power. Incited by the persecution of Protestants by the Spanish Inquisition, in 1566 Calvinist zealots had started a series of iconoclastic attacks, during which mobs defaced and invaded churches to destroy statues and organs, decorations and fittings. Philip II, the Catholic king of Spain and Portugal and ruler of the Netherlands, reacted with brutal repression, only to fuel the conflict that eventually escalated into the Eighty Years' War, also known as the Dutch War of Independence. In 1581 the country, being the most densely populated and productive region of the Spanish Empire, declared its independence and became a safe haven for refugees, among them many skilled Protestant craftsmen and wealthy merchants from the southern Netherlands (today's Belgium) still under Spanish control, but also Sephardic Jews from Spain and Portugal. Jews living in the reconquered Islamic territories had been first forced to convert and then persecuted. In public autos-da-fé that competed with bullfights for the public's attention, tens of thousands had been murdered and hundreds of thousands tortured by the Inquisition. These well-educated refugees established a climate of intellectual tolerance in the cosmopolitan cities that propelled the country into the "Dutch Golden Age." Holland, traditionally home to the world's best mapmakers, developed into an unparalleled hot spot for commerce, science, and the arts. Dutch merchants established the largest corporation in history, financed by shares traded at the first modern stock exchange, Dutch colonists founded Nieuw Amsterdam, the later New York on the southern tip of Manhattan Island, and Dutch painters produced millions of masterpieces. At the same time, Christiaan Huygens realized that what Galilei called Saturn's "arms" were actually rings, discovered the law of momentum conservation and the wave theory of light, and invented the magic lantern and the pendulum clock. There were downsides to these glorious developments, of course, as a significant part of the wealth was generated with the exploitation of indigenous cultures and slave trade. Also, republican patricians quarreled for power with the princes of Orange, and the far-reaching freedom of opinion and speech led to parallel societies and increasing conflicts between progressive liberals and Calvinist zealots. It was this intellectual climate that made possible a philosopher who was a champion of both monism and rational thinking, too radical even for the tolerant Amsterdam.

Baruch Spinoza was part of the community of Sephardic Jews in Amsterdam, where his grandfather, who had been fleeing from Portugal, had finally found asylum, and he has been praised for being among the boldest thinkers ever—not only because he broke with the traditional views of his own and essentially all other religions, resulting in the harshest ban ever pronounced by the Sephardic community of Amsterdam, but even more so since he dared to try to understand the entire universe in terms of a philosophical "theory of everything," based exclusively on rational thought.[44] In Spinoza's universe, everything that is, is necessary and determined; there are neither miracles nor contingencies. Even God has no choice about what to do with the universe; nor did he even exist before the world: "Since there is neither inconstancy nor change in God, He must have decreed from eternity that He would produce those things which He produces now," Spinoza wrote. "It follows that all created things have been under an eternal necessity to be in existence." He added that "since in eternity there is no when, or before, or after, or any other change of time, it follows that God did not exist before those things were decreed, to be able at all to decree otherwise."[45] Thus, according to Spinoza, the universe itself is a single, infinite, eternal, necessarily existing substance to which the terms "God" and "nature" apply equally well. In his *Short Treatise on God, Man and His Well-Being*, Spinoza redefines God's providence as "nothing else than the striving which we find in the whole of nature and in individual things to maintain and preserve their own existence."[46] In his *Ethics*, published shortly after his death, he is even more unabashed, speaking of "the eternal and infinite Being, which we call God or Nature."[47] Spinoza denied the immortality of the soul, rejected the notion of a providential God, and denied that the commandments of the Torah were literally given by God and still binding for Jews. For most of his contemporaries, Spinoza was simply an atheist, since he explicitly identified nature and God, and his God aka nature was decidedly monistic: "The whole of Nature is but one only substance . . . , all things are united through Nature, and they are united into one [being], namely, God."[48]

To be accurate, though, Spinoza distinguished two kinds of nature: *natura naturata* ("natured" nature), what John Scotus Eriugena had referred to as *natura creata* and a proper characterization of quantum mechanics' on-screen reality, and *natura naturans* ("naturing" nature), what Eriugena had called *natura creans* and a suiting description for the projector reality of quantum mechanics. "In other words," as Bernard d'Espagnat illustrated in his *Veiled Reality*, this quantum concept "plays, in a way, the role of the God—or Substance—of Spinoza, although there are some differences."[49] Among the two realities, the projector reality is the foundational one: "Substance is, by its nature, prior to all its modifications," Spinoza explains.[50]

The modest philosopher, who never held an academic position (he had declined an offer from the University of Heidelberg, intent on maintaining his independence), became a leading figure of Enlightenment philosophy. Jonathan Israel, professor emeritus at the Institute for Advanced Study, credits Spinoza as "the supreme philosophical bogeyman of Early Enlightenment Europe."[51] From his private home, Spinoza communicated with and inspired the father of modern chemistry, Robert Boyle, physicist and mathematician Christiaan Huygens, and philosopher Gottfried W. Leibniz. There he was visited by the designated secretary of the Royal Society, Henry Oldenburg, who later expressed his hope that Spinoza and Boyle would unite their abilities "to advance a genuine and firmly based philosophy."[52] Later both Erwin Schrödinger and Albert Einstein were deeply inspired by Spinoza. As

Max Jammer has pointed out, Einstein had been reading Spinoza already as a student in Berne, and his philosophical views were "near to those of Spinoza."[53] In his appreciation, Einstein even went as far as to compose a poem of praise: "How much do I love that noble man, More than I could tell with words, I fear though he'll remain alone, With a holy halo of his own."[54] For "the holy excommunicated Spinoza," as the German theologian Friedrich Schleiermacher described him, "the universe [was] his only and lasting love."[55]

As it turned out, Spinoza may have been too radical even for the liberal Netherlands, while the Dutch Golden Age collapsed under the strain of the nation's inner conflicts. When in 1672 the Netherlands were drawn into a war with both France and England, these conflicts led to the public murder of Johan de Witt, a wealthy republican who had fostered Spinoza's work and the leader of the States of Holland for the past twenty years, and his brother Cornelis. When Johan tried to pick up his brother after he had been tried for treason, corruption, and atheism and sentenced to exile, both brothers were attacked and murdered by a fanatic mob, presumably with the approval of the prince of Orange, William III, in front of the prison in the middle of the street. "They stripped the bodies bare, cut off . . . the genitals, slit the bodies open and pulled out the hearts and entrails . . . A few participants . . . even roasted parts of the cadavers and ate them," historian Herbert H. Rowen recounted.[56] The year was 1672, known in Dutch as the "Disaster Year," which signified the end of the Golden Age. The country survived, and William III, as its new national leader, even became the king of England; yet the impetus of power and progress moved on to England and France.

In the aftermath of the Disaster Year, Spinoza's books were forbidden, though his philosophy remained influential. Spinoza's correspondent Henry Oldenburg, the first secretary of the Royal Society, disseminated it via the extensive network of scientific contacts he had established to stimulate the exchange of Europe's best minds. It was Oldenburg who founded the *Philosophical Transactions of the Royal Society*, the first journal dedicated exclusively to science, and installed the process of peer review to scrutinize the academic quality of contributions by fellow scientists, to make science more transparent, collaborative, and modern. In this spirit, Oldenburg also contacted the unknown Isaac Newton, persuaded him to join the Royal Society, convinced him to publish about the invention of the Newtonian telescope, and initiated the discussion of Newton's work with other scientists such as Huygens and Robert Hooke. As a consequence, Newton's work unleashed a new revolution in science.

A Clockwork Universe, Powered by the Pipes of Pan

Isaac Newton is usually considered the herald of modernity and "chief architect of the modern world," as his biographer James Gleick describes him.[57] In his magnum opus, *Philosophiae naturalis principia mathematica*, credited as one of the most important works in the history of science, Newton derived Kepler's laws of planetary motion, developed his law of universal gravitation, and established the bedrock of classical mechanics so exhaustively that the discipline is often simply referred to as "Newtonian mechanics," a system of the world that is so rational and deterministic that it is customarily compared to clockwork. Through Newton, according to cosmologist Hermann Bondi, "the landscape has been so totally changed, the ways of thinking have been so deeply affected, that it is very hard to get hold of what it was like before."[58] When the English Romanticist William Wordsworth portrayed Newton as an ultrarational, dispassionate loner, "with his prism and silent face, The marble index of a mind forever Voyaging through strange seas of thought, alone," this in fact describes fittingly the small but ambitious, lonely farm boy, who observed attentively the shadows moving along the walls of his grandparents' house and then finally hammered wooden pegs into the stonework to construct a sundial and measure the time from the cycles of the sun. There was another side to Newton though, a Newton who "was not the first of the age of reason," as economist John Maynard Keynes has pointed out, but "the last of the magicians"—a Newton who dabbled in biblical exegesis and alchemy and who, according to historians of science James McGuire and Piyo Rattansi, pursued these endeavors "in as rigorous a fashion as his scientific work."[59]

Both of these seemingly conflicting traits can be traced back to a deeply rooted monistic conviction. As McGuire and Rattansi have pointed out, Newton had composed extensive notes on an ancient, monistic underpinning of his physics that he planned to include in the revised version of his *Principia*: "The sheer bulk of the manuscripts, the number of copies and variants, their relation to Newton's other writings, and the testimony of Newton's associates . . . all make it certain that he considered the arguments and conclusions . . . an important part of his philosophy."[60] A crucial source of inspiration for this background was Newton's older colleague at the University of Cambridge, Ralph Cudworth.

In England, the Renaissance Platonism had originally manifested itself most prominently in poetry. In the footsteps of Florentine poet Girolamo Benivieni, a member of Ficino's academy who had composed a poetic interpretation of Ficino's interpretation of Platonic love, describing "how Love from its celestial source, . . . flows to the world of sense, . . . moves the heavens, refines the soul, gives laws, . . . And from that soul, . . . Is earthly Venus born, whose beauty lights, The skies, inhabits earth, is nature's veil," these ideas became a powerful influence on love lyrics, first in Italy and later in England.[61] Edmund Spenser's poems *Fowre Hymnes* and *The Faerie Queene*—in which nature is personified by a woman with a veil, for example—exhibit a distinctly Platonic flavor, as does his friend Philip Sidney's *Astrophil and Stella*, which influenced William Shakespeare. Sidney also was a close friend of Giordano Bruno's during his stay in England, where Bruno used Ficino's books in his lectures. In the seventeenth century, these traditions inspired a group of philosophers who became known as the Cambridge Platonists. A prominent member of this group was Ralph Cudworth, who was deeply worried about the increasing popularity of what he perceived as materialism and atheism, including most notably

Spinoza's monism. In his magnum opus, *The True Intellectual System of the Universe*, from 1678, Cudworth thus developed his own version of monism to rebut these philosophical trends. In an ironic twist of history, Cudworth's work was later confused with the philosophy of Spinoza, which he was trying to refute with a somewhat odd argument. Based on Ficino and Pico della Mirandola, Cudworth argued that all belief systems—including atheism and materialism—could actually be traced back to a common source: a primordial, monistic *prisca theologia* or *philosophia perennis* that had once been revealed by God and shared by the ancient Egyptian religion, prophets, and sages, such as Moses, Orpheus, Pythagoras, and Plato, and had been watered down ever since. As the core concept of this philosophy, Cudworth identified the monistic *Hen Kai Pan*, which is Greek for "One and All."

Little known is that Newton strove to rediscover this distinctive monistic philosophy in both his physics and his nonscientific activities in alchemy and biblical exegesis. "Newton believed that he knew how God's agency operated in His created world," as McGuire and Rattansi write, and just like the Cambridge Platonists, he was convinced he would rediscover the ancient knowledge of this *prisca theologia*.[62] As McGuire and Rattansi proved, Newton "saw the task of natural philosophy as the restoration of the knowledge of the complete system of the cosmos."[63] Parallels to the Renaissance philosophy of Ficino and Pico della Mirandola, as well as the thought of Giordano Bruno, are obvious and striking. Indeed, Newton's notes are full of references to Plato, Pythagoras, and the music of the spheres, which he interprets as an allegory for gravity. McGuire and Rattansi determine a "strongly Platonic bias in his authorities."[64] About the various planets, elements, and phenomena that were deified in antiquity, Newton explains, "these things All are one thing, though there be many names," in other words, "one and the same divinity exercising its powers in all bodies whatsoever," as Newton approvingly explains the monistic philosophy of antiquity.[65] Likewise, in a draft of his *Opticks*, Newton wondered, "What is it, by means of wch [sic], bodies act on one another at a distance. And to what Agent did the Ancients attribute the gravity of their atoms and what did they mean by calling God a harmony and comparing him & matter to the God Pan and his Pipe."[66] Pan, as Newton explains in his notes, "was the supreme divinity inspiring this world with harmonic ratio like a musical instrument" and struck "the harmony of the world in playful song."[67] As McGuire and Rattansi have pointed out, to Newton the universal gravitation that, according to his famous discovery, is responsible for both objects on Earth falling down and the celestial pull that forces the moon to orbit Earth, and that could be generalized to the planets and finally to all celestial objects whatsoever, was inseparably linked to his belief that "gravity was a direct result of the exercise of divine power."[68] This all-encompassing divine power permeated everything that is, Newton believed, and has particularly interesting parallels to the recent research aiming to derive gravity and space-time from entanglement (see Chapter 7).

As McGuire and Rattansi have pointed out there was no "multiplicity of Newtons" that can be characterized as either "the last of the magicians" or "the first of the scientists."[69] Instead, the historians of science find sufficient evidence that Newton's entire edifice of ideas is permeated by monistic philosophy. "The mechanics and cosmology of the Principia [were] influenced by Newton's theological views and his belief in a pristine knowledge"—and thus by his monistic philosophy—and these topics constituted "for him, as for many of his contemporaries, ... the ultimate problems."[70] We will find this pattern recurs in future scientific breakthroughs until the present day.

Conspiracy and Poetry

Despite how little known the influence of monism on the development of science is, this process wasn't over yet with Newton and his discovery of classical mechanics. It was the Romantic Science of the Second Scientific Revolution that developed concepts such as field theory that seem to lie at the heart of the challenges modern physics faces today. At the same time though, Romanticists came up with spoiled, subjective notions that eventually reinforced the association of monism with pseudoscience. While in the course of the eighteenth century Newton's clockwork universe had been generally accepted, where it came from was forgotten. Monistic philosophy made another detour through politics, secret societies, and poetry, before it was fed back into science.

In the early 1700s, the Irish philosopher John Toland, a contemporary of Newton, had added a new, political twist to the speculation about a monistic, primordial truth. Toland had started out as a deist, arguing that Christianity could be understood exclusively by reason, only to see his books being burned publicly in Dublin. Around the same time, in 1697 Thomas Aikenhead, a twenty-year-old student at the University of Edinburgh where Toland had achieved his master's degree, was hanged for blasphemy after he had claimed that "God, the world, and nature, are but one thing."[71] It must have been shocking for Toland to realize that such an act of injustice still could happen, and not in Inquisition-oppressed southern Europe but in Presbyterian Scotland.

After these incidents, Toland became increasingly radical. In his *Letters to Serena*, according to Toland's biographer Justin Champion, "conceived as an antidote to the damage false religion and superstition did to civic communities" and addressed to Sophia Charlotte of Hanover, the later queen of Prussia and sister of King George I, ancestor of the British royal family, Toland was the first to coin the word "pantheism" and to describe himself as a "pantheist."[72] As historian of science Margaret Jacob explains, Toland used "Newton's science to argue that motion is inherent in matter" or "in other words, that nature can govern itself, that motion, life, and change have entirely naturalist explanations."[73] Following in the footsteps of earlier monists, Toland identified the Egyptian goddess Isis as "Nature, the parent of all things."[74]

Toland was also involved in the circulation and probably also the composition of a dangerous, clandestine text describing the monotheistic religions as fraud. Drawing on Lucretius, Bruno, Spinoza, Lucilio Vanini, and other heretics, Toland dismissed priests as agents of tyranny and, more than a century before Karl Marx, spoke of religion as the "opium of the people."[75] Moses, Jesus, and Muhammad are depicted as "impostors": "false prophets who manipulated religion to their own end."[76]

What's new is that Toland blends his monism with (in his time radical) political liberalism and (even by today's standards radical) anticlericalism. Toland advocates liberty, religious tolerance, and republicanism, including full citizenship and equal rights for Jewish people, and becomes an activist for Whiggism, the liberal political philosophy that opposed a Catholic monarch and advocated a supremacy of the parliament over the king in the aftermath of the English Revolution. Following Spinoza and Toland, monism got associated not only with heresy but increasingly also with progressive politics.

In his last work, *Pantheisticon*, Toland developed a liturgy for pantheists that has been likened to the rituals of the increasingly popular lodges of Freemasons and asserted that "true knowledge of nature would liberate the mind from the dark shadows of superstition and consequently dissolve the grounds of political tyranny," as Justin Champion writes.[77] Indeed, it has been speculated that Toland was affiliated with an early Masonic group, and it is generally accepted now that at least some Masonic lodges—besides engaging in obscure rituals and networking (some would say wangling)—by the circulation of clandestine literature, by cultivating a libertine tone, and by providing safe spaces for free thinking and speech also became powerful catalysts in spreading Enlightenment principles.[78] Many of the early English Freemasons were French Protestant refugees and Whigs, with excellent connections to the natural philosophers of the Royal Society. Indeed, all of the secretaries of the Royal Society from 1714 to 1747 were Freemasons. A particularly famous Mason was John Theophilus Desaguliers, member of the Royal Society and experimental assistant to Isaac Newton, who wrote a poem to praise a blend of Newtonian philosophy, Masonic symbols, and Hanoverian monarchy as universal harmony. Another refugee of French origin was John Coustos, who had made it back to London after he had been kidnapped, tortured, and threatened with being burned alive by the Portuguese Inquisition. Coustos's book describing this misery became a best seller and added to the appeal of Freemasonry, even though (or because) he exaggerated both his torture and his bravery in not betraying the Masonic secrets. In the following years, Freemasonry developed into an international trend, with secrecy and the claim to be in possession of arcane truths as its trademarks. Pythagorean secrecy turned out to be a perfect match for Masonic rituals, and few philosophies would have been better suited as an integral part of Masonic mythology than the *prisca theologia* or *philosophia perennis* conceived of by Ficino, Pico della Mirandola, and Cudworth and searched for by Newton.

So, for much of the eighteenth century, monism remained hidden behind the closed doors of such secret brotherhoods. That changed when at least one Masonic group in Germany turned its Enlightenment agenda into political activism. By 1780, the Bavarian Illuminati had started to successfully infiltrate government administrations with the ultimate goal of overcoming the dominion of people over people and establishing a free society. When its members engaged in espionage and document theft, the order was outlawed around 1785. Only four years later the breakout of the French Revolution shocked absolutistic rulers throughout Europe, and the brief existence of the Bavarian Illuminati inspired various conspiracy theories that vilified this and other Masonic groups and accused them of instigating revolutions and pursuing universal domination within a "new world order." In this emotionally charged climate, the conservative scholar Friedrich Heinrich Jacobi provoked an argument that uncovered the widespread pantheism among leading German intellectuals such as Gotthold Ephraim Lessing, Johann Gottfried Herder, and Johann Wolfgang von Goethe.

Goethe, the polymath and literary genius who became a German national idol, had harbored pantheistic convictions from an early age. In his autobiography, he reminisced about how as a young boy he almost set his father's furniture on fire when he tried to mount a flame atop an altar he had assembled with a collection of natural objects, presumably leaves, feathers, fossils, and interesting-looking stones and flowers. These items should represent the world and worship of "the great God of nature" through his creation, the God "of the motion of the stars, the days and seasons, the animals and plants," as Goethe writes.[79] Appropriately, when the young poet attended the university in Strasbourg and fell in love, he turned to nature to express his feelings: "How fair doth Nature Appear again! How bright the sunbeams! How smiles the plain!" Goethe extols.[80] Around the same time, Goethe composed the poem *Prometheus* glorifying the mythological Titan's rebellion against Zeus, a barely concealed attack against an authoritarian, monotheistic God. The inflammatory work didn't get published until fifteen years later, when Goethe's friend Friedrich Jacobi—without the author's knowledge or consent—took advantage of it in his attack against Spinozean philosophy. Five years earlier, in 1780, Jacobi had visited his fellow Freemason, the famous writer and Enlightenment scholar Gotthold Ephraim Lessing, and showed him Goethe's poem. To Jacobi's consternation, Lessing wasn't repelled but admitted that he himself was a pantheist: "The orthodox concepts of divinity are no longer for me; I cannot enjoy them. One and all! I know nothing else."[81] After Lessing died one year later, Jacobi denounced Lessing's pantheism and caused a scandal that propelled a major shift in German philosophy. Ironically, contrary to Jacobi's objective, this episode, known as the "pantheism controversy," triggered a substantial and lasting Spinoza revival.

One famous reader who came under the spell of Goethe's *Prometheus* was the voyager, naturalist, and later revolutionary Georg Forster (1754–1794), who took part as a teenager in James Cook's second circumnavigation from 1772 to 1775, a trip that imbued him with monistic spirit: "I am indebted to that voyage for the development of an endowment, which determined my course in life from childhood on, namely, endeavoring to trace my ideas back to a certain universality, bundling them into a unity, and thus endowing an awareness of the whole of nature."[82] Just like his contemporaries Goethe and the young Alexander von Humboldt, who had accompanied Forster on several journeys and for whom he became a role model, Forster was convinced that "everything is connected through the finest modulations," that "the order of nature does not follow our divisions."[83] Forster is also an exemplar of the link between eighteenth-century monism and progressive politics: when the French Revolutionary Army conquered his hometown, Forster joined the revolutionaries and worked to establish the first democratic state on German territory, the short-lived Republic of Mainz. It lasted only four months, until Mainz was recaptured by the Prussian army. Forster, who had been sent for negotiations to France, died only one year later in Paris, lonely and impoverished.

Meanwhile, monistic philosophy was becoming increasingly popular. By the end of the eighteenth century, monism, now associated with arcane truths and progressive politics, had made it back into mainstream culture and became a powerful motif in the arts. References to monistic motifs can be found in the US Declaration of Independence with its invocations of the "powers of the earth," the "Laws of Nature," and "Nature's God"; in Friedrich Schiller's "Ode to Joy," extolling how "magic binds again" what custom had strictly divided, how "at nature's breasts" drink all the just, all the evil, to follow her trail of roses and dedicating "this kiss to all the world!"; or in the invocation of the Egyptian goddess Isis in Wolfgang Amadeus Mozart's opera *The Magic Flute*. In fact, Egyptologist Jan Assmann diagnoses a "mystery fever" and a downright "Egyptomania" in the last quarter of the eighteenth century: a "flood of texts about ancient mystery cults that never before or after have attracted a comparable attention."[84] Ludwig van Beethoven, who set Schiller's "Ode to Joy" to music in the version that was chosen later as the anthem of Europe, kept the sentence "I am all that is," copied from Schiller's essay "The Mission of Moses" and attributed to Isis as an embodiment of monism, in a frame on his desk until he died in 1827.[85] For the profoundly influential Goethe, monism remains a recurring theme in his work throughout his life, from his 1784 scientific essay "On Granite," in which he emphasizes that "all natural things are related with each other"; to his poems "Reunited" and "One and

All," in which he—in the tradition of Plotinus, Eriugena, and Ficino—describes the creation of the universe as a separation from God that can be vanquished in love and how "the eternal bestirs itself in everything"; to his *Xenia* from 1827, in which he writes that "Isis shows herself without a veil, but man has a cataract." Also, the German edition of naturalist and explorer Alexander von Humboldt's "Essay on the Geography of Plants" credits Goethe's inspiration with a dedication and an engraving that shows the genius of poetry unveiling Isis.

While Goethe is mainly famous for his poetry, he took a keen interest in science and also performed scientific studies himself. Independently of the French physician Félix Vicq d'Azyr, he discovered, for example, that human embryos possess an incisive bone just like those of other mammals, a finding that provided the missing link between humans and animals and could be interpreted as evidence for biological evolution. It was through Goethe and the Romanticists that monism eventually got fed back into philosophy and science and resurfaced as a major source of inspiration in the Second Scientific Revolution. Among the key figures in this process was a young prodigy, who was still a teenager in the 1790s but would go far.

"Hen Kai Pan"

"Hen Kai Pan!—One and All!" For Friedrich Schelling and his friends, the strange-sounding, muttered words were all one at once: codeword, greeting, and motto, the phrase literally meant the world. Schelling had been only fifteen years old when he enrolled in the Tübingen Stift and was five years younger than his roommates, the sensitive and original Friedrich Hölderlin and the cumbersome, earnest, brooding Gottfried Wilhelm Friedrich Hegel, who already in his youth was known by the nickname "Old Man." By then, Schelling had already taught himself eight languages, including Hebrew and Arabic. While all three of them were highly gifted, and all of them were to become famous, none of them was to become a priest. Together, they suffered from the strict rules in this "galley of theology," as they described it in desperate letters to their parents. Yet the Stift was also considered to be the most prestigious seminary for the education of future Protestant priests in the country. There, exactly two hundred years before, Johannes Kepler had been educated. Dreaming up to the skies, Hölderlin had composed a poem to celebrate the man who had made sense out of the planetary orbits around the sun, interpreted them as a cosmic harmony, and inspired Newton's classical mechanics: "In starry regions my mind perambulates, hovers over Uranian fields, and ponders. Solitary and daring is my course, demanding a brazen stride... Who led the thinker in Albion..., into the field of deeper contemplation, and who, lighting the way, ventured into the labyrinth."[86]

Like many young minds of their generation, the three friends felt restless. The lack of individuality and freedom in the Stift only reflected the political state of affairs outside. In this situation, their slogan evoked a secret paradise—and in the earthshaking years following 1789, there were good reasons for secrecy. Sure, Tübingen was a quiet, small town with half-timbered houses and a medieval castle in its center, overlooking lush green hills and the outskirts of the Swabian Jura. But Tübingen was only seventy miles from Strasbourg and the French border, a distance a fast horse could travel in a few hours from the country where the French Revolution was powdering Europe's feudal order into pieces.

When the friends joined a group of students to plant a liberty pole on a meadow outside the town and signed a pamphlet with revolutionary slogans such as "Vive la liberté!" the Duke of Württemberg Carl Eugen came in person to the Stift to investigate the case. To the friends' relief, the only consequence was that the group's leader had to escape to Strasbourg. Meanwhile, the three friends' curiosity wasn't confined to politics. They corporately "sacrificed Bacchus" and joined for heavy drinking. And they began to get up in the middle of the night—two hours before the daily curriculum started—to discuss philosophy. In hindsight, these discussions were just as revolutionary as the politics. In 1781 Immanuel Kant had published his *Critique of Pure Reason*, in which he had pointed out the impossibility of accessing reality as it really is. For Kant, the "Thing-in-itself," the unconditioned absolute, remained elusive. Instead, we always experience reality indirectly, through our senses. Space and time, for example, were described by Kant as a pair of glasses through which we perceive the world. Johann Gottlieb Fichte had taken this view to the extreme by advocating that it is the subject who creates reality—a conception that comes close to John Wheeler's "participatory universe," illustrated by his enigmatic "U," two hundred years later. As it turned out, such a view fit snugly with the upcoming intellectual movement of Romanticism, which placed the creative ego up front.

Romanticism represented a rebellion of the individual against many things: against a state and society where everything was decided by nobility, kings, and queens; against a paradigm of nature working as a mechanical clockwork; against a boring and gray daily life that left no space to follow one's dreams. Sparked by the French Revolution's demand for freedom for the individual and by Fichte's philosophy emphasizing the role of the individual in the emergence of our experience, Romanticism became the predominant intellectual movement in the early nineteenth century. The Romanticists idealized nature and arts, where they sought what they were lacking in their daily routines. In nature, the Romantic ego hoped to recover itself. But to find the ego in nature, ego and nature needed to have a common source. The Romanticists needed a principle to reconcile ego and universe, mind and matter, thought and desire. They were in need of "Hen," of an all-embracing "One," and the three friends became pivotal in supplying it.

The creative Friedrich Hölderlin turned to poetry and wrote hymnic appraisals of becoming one with nature: "To be one with all—this is the life divine, this is man's heaven . . . to return in blessed self-forgetfulness into the All of Nature—this is the pinnacle of thoughts and joys, this is the sacred mountain peak, the place of eternal rest."[87] Hölderlin became one of the most important poets of his time, until, at only thirty-five, he was diagnosed with incurable mental illness and spent the rest of his life in a tower by the Neckar riverbank, just a few steps away from the Tübingen Stift. Hegel and Schelling would devote their lives to striving for the true philosophy of "Hen," first as comrades and later as bitter rivals. Schelling in particular became a fast-rising star and developed into the prime philosopher of the Romantic movement.

In 1798, at the age of twenty-three, he was hired by Goethe as a professor in Jena, the university of the liberal duchy around Weimar, known as "Athens at the River Ilm." In Jena Schiller and Fichte were already teaching, until the latter was fired for atheism. Yet Schelling's monistic philosophy resonated even better with the convictions of Goethe, who himself wrote about the unity of nature in a way closely reminiscent of Hölderlin. No wonder Schelling soon became part of Goethe's inner circle and got acquainted with the poets of German Romanticism.

In an unpublished poem from around this time, Schelling demanded that a true religion should exhibit itself in stones and moss, flowers, metals, and everything, so that one may immerse oneself "in the universe" or "in one's lover's fair eyes."[88] Yet the lover he was referring to was married to another man. Caroline, wife of Romanticist August Schlegel, would be the love of his life. Twelve years his senior, she was intelligent, highly educated, and emancipated—and had already lived an adventurous life: at the age of thirty-six, she had—together with her husband—translated six of Shakespeare's plays into German, given birth to four children (one of them the son of a French revolutionist officer with whom she had engaged in a one-night stand during a ball), and been widowed and remarried. When living in the Republic of Mainz, she had been close friends with Georg Forster, who by then served as the city-state's vice president. When the Prussian troops recaptured the city, she had been imprisoned and, as a "democrat" and "lady of easy virtue," banned from many places, including her hometown of Göttingen. After she married Schlegel, the couple moved to Jena, where their household became a Romanticist hub.

It was then that she met Schelling, whom she described as a "primeval figure" and compared with "granite." Soon Schelling and

Caroline began an affair that was tolerated by her husband. The years in Jena were probably the happiest in Schelling's life. With the help of Goethe, he got Caroline to obtain a divorce and marry him, but the couple would live together for only six years. When Caroline died in 1809, Schelling felt that the last tie connecting him with the world was cut, and his philosophy acquired a dark note, speaking of chaos, accidents, and an irrational desire at the foundation of all things. But these times were still in the future when, in 1799, Schelling got Hegel a position in Jena, and the two old friends worked so closely together that it is sometimes hard to reconstruct who wrote what. In these years Schelling developed his "Naturphilosophie," a philosophy of nature that would shape the science of his time and the development of modern physics.

To a significant degree, Schelling's philosophy reflected his conflicted personality. He was a man torn between inwardness and extroversion, between the absolute and the concrete, between spirit and nature, between religion and atheism. Throughout his life Schelling struggled to reconcile Fichte with Spinoza, ego with the world, and an all-encompassing absolute with the diversity experienced in nature. Adopting Spinoza's monism, Schelling tried to unify matter and mind: to him, everything was godlike, from the crystal to the leaf and from the leaf to human nature. And just as Heraclitus wrote both "from all things One and from One all things" and "war is the father of all," Schelling's philosophy reflects the ideas of Georg Forster and circles around the principles of unity and of opposing polarities complementing each other to form an integrated whole: man and woman, electricity and magnetism.

When Niels Bohr's philosophical background is linked to German idealism, this refers to the philosophy of Kant, Fichte, Hegel, and Schelling. In fact, Bohr's principle of "complementarity," the idea that contrary descriptions such as "particle" or "wave" are not contradictory but complementary, strongly resembles Schelling's opposing polarities, and Bohr's emphasis on the importance of the observer recalls the Romanticists' creative ego.

But this story tells only half the truth. In contrast to Schelling, Bohr was anything but a monist. In fact, as we have seen before, it was a defining element of Bohr's philosophy to deny the entity that corresponds to the combined complements as part of reality. Thus, if Bohr really adopted the philosophy of German idealism, he strongly blended it with positivism (i.e., the view that only that which is observable is real). In contrast, according to Michela Massimi, a philosopher at the University of Edinburgh, "Schelling's main contribution was to naturalize Kant's notion of the unconditioned and to transform it into an object of empirical investigation for Naturphilosophie [i.e., philosophy of nature]. In doing so, Schelling took Kant's important insight about the perspectival nature of human knowledge . . . and turned it upside down."[89] As Schelling put it, "Empiricism extended to include unconditionedness is precisely philosophy of nature."[90]

Schelling thus complements the experimental investigation of nature with theoretical model building, allowing for abstract conceptions that are not directly observable but still part of reality and even more real than empirical knowledge, which always is subject to a specific perspective. In doing so, Schelling developed concepts that exhibit astonishing parallels with quantum mechanics. In his *First Outline of a System of the Philosophy of Nature*, Schelling introduced "dynamic atoms" at the foundation of nature, which do not exist in space and "cannot be viewed as part of matter"[91] but are rather "constituent factors of matter"[92] whose effects and products "are presentable in space."[93] The dynamic atom itself "is nothing other than the product itself viewed from a higher perspective."[94]

The similarity with quantum objects that describe physics before measurement in abstract Hilbert spaces (as illustrated in Chapter 1 with Plato's Allegory of the Cave) is startling—one hundred years before scientists even started to explore the physics of the quantum realm. As Massimi describes it, "Schelling boldly relocated [Kant's] unconditioned in the realm of theoretical reason, in the form of the scientific knowledge of nature."[95] In addition, Schelling's monism provides a surprisingly accurate description of entanglement. And finally, in his struggle to understand how many things can originate from "One," Schelling, more concretely than Goethe before him, echoed Eriugena and anticipated the basic idea of decoherence. When he explains that "before, humans lived in a natural state, [and] man was one with the surrounding world," this still sounds like a vague commitment to primeval monism, but when he goes on to explain how "as soon as a human sets itself apart, in opposition to the outside world . . . he separates what Nature had unified, he separates object from observer . . . and finally (by observing himself) himself from himself," this boils down to an amazingly accurate description of the quantum factorization into observer, observed system, and environment, thereby giving rise to the decohered frog perspective.[96] Indeed, for Schelling, philosophy (or science, that is) adopts a "God's perspective," understanding the totality of nature as a whole. And according to his friend and collaborator Hegel, what we observe in nature is God himself, although perceived as something "alien," as seen from outside instead of from within.

Obviously, Schelling and Hegel shared similar thoughts with Hölderlin, who gave them a more poetic expression: "But an instant of reflection hurls me down. I reflect, and find myself as I was before—alone, with all the griefs of mortality, and my heart's refuge, the world in its eternal oneness, is gone; Nature closes her arms, and I stand like an alien before her and do not understand her."[97]

From Frankenstein to Fields

Schelling's career paralleled the victory march of Romanticism: he received a knighthood; the king of Prussia wooed him, as "the God chosen teacher of the time," to join Berlin University; eventually the king of Bavaria engraved "to Germany's first thinker" on his tombstone. After all was said and done, "Pantheism is the clandestine religion of Germany," as the poet Heinrich Heine would declare only a few years later, and Schelling's influence would spread far and wide.[98]

After the English poets Samuel Taylor Coleridge and William Wordsworth visited Germany in 1798, they carried Romanticism and a monistic apprehension of nature back to Britain, where they set out to write a poem to change the world. While this poem was never accomplished, in his works Wordsworth would speak of himself as "a worshipper of Nature"[99] and of nature as "the breath of God."[100] These ideas would inspire the art of William Turner and the French Impressionists, drive Claude Monet, Auguste Renoir, and Vincent van Gogh to leave their studios and paint in nature, and impel scientists to leave their laboratories. In this spirit Alexander von Humboldt set out to ascend Chimborazo, thought to be the highest mountain on Earth, trying to document and measure everything, while conceiving nature as a "reflection of the whole" and a "net-like intricate fabric."[101] The monistic glorification of nature would become an integral component of the "frontier myth" that looked at the great wilderness of the American West as a universal promise, conveyed through the works of Walt Whitman, Ralph Waldo Emerson, Henry David Thoreau, John Muir, Ansel Adams, Jack Kerouac, and many others. A wonderful expression of this sense of an all-encompassing nature is Whitman's poem titled with the Greek-German word "Kosmos": "Who includes diversity and is Nature, . . . The past, the future, dwelling there, like space, inseparable together."[102] Finally, a monistic cosmos would become once more a driving force for science, and in England monism-inspired physics would undergo its next revolution.

When in the early nineteenth century scientists had started to investigate phenomena such as electricity and magnetism, heat, fluidity, and life, the limitations of explanations within the context of Newton's classical mechanics became more and more obvious. In this situation, the Enlightenment concept of a clockwork universe appeared increasingly unsuitable. In its place came a revival of the old allegory of the universe as an organism.

In this situation, Schelling's monistic philosophy became a major source of inspiration for the early pioneers in electrochemistry and electromagnetism, who had had close ties with the intellectual trend of Romanticism. Defining elements of this philosophy were the idea of unification, energy conservation, and the balance of opposing polarities complementing each other, all melded within a monistic perspective onto the universe. According to the late American philosopher of science Thomas S. Kuhn, "Schelling ... maintained 'that magnetic, electrical, chemical and finally even organic phenomena would be interwoven into one great association ... [which] extends over the whole of nature.'"[103] This had concrete implications for how the new phenomena of electricity and magnetism were approached, as Kuhn explains: "Even before the discovery of the battery [Schelling] insisted that 'without doubt only a single force in its various guises is manifest in [the phenomena of] light, electricity and so forth.'"[104] Indeed, as Kuhn pointed out, "many of the discoverers of energy conservation were deeply predisposed to see a single, indestructible force at the root of all natural phenomena."[105] Obviously, Kuhn concludes, many natural philosophers in Schelling's tradition "drew from their philosophy a view of physical processes very close to that which Faraday and Grove seem have to drawn from the new discoveries of the nineteenth century."[106] Giles Whiteley, an expert on Schelling's reception in nineteenth-century British literature, agrees: "It was more or less directly under the inspiration of Schelling that Ritter developed the field of electrochemistry, and it was through combining the discoveries of Davy and Ritter that Hans Christian Ørsted discovered the principle of electromagnetism in 1820."[107]

Johann Wilhelm Ritter (1776–1810) was a physicist in Goethe's circle, who discovered ultraviolet radiation and the rechargeable battery. Ritter's research on "galvanism," however—the idea of creating life from inanimate matter by means of electricity, sparked by the observation that electric currents could induce muscle contractions in frog legs or executed criminals—finally led him astray. After he had relocated to Munich, he became increasingly obsessed with his research, neglected his family, moved into his laboratory, and died young and poor in 1810, arguably mainly as a result of his self-experimentation with electricity. As historian of science Richard Holmes has speculated, Ritter may have been the role model for Dr. Frankenstein, the title character of the novel young Mary Shelley wrote six years later when she, her stepsister Claire Clairmont, and her lover and later husband, the poet Percy Shelley, spent the "wet, ungenial summer" of 1816 visiting the famous Romanticist Lord Byron, who was both Claire's lover and stepbrother.[108] After Byron had been forced to leave England by his increasing debt, the scandal about the split-up with his wife, who just had given birth to a baby daughter and believed her husband had gone mad, and the rumors about his incestuous affair with Claire, he was living now in a mansion near Lake Geneva. The year 1816 became known as the "year without summer," caused by volcanic ash from a massive eruption (the largest one for thirteen hundred years) of an Indonesian volcano the year before. Consequences included floods, brown snow falling throughout the year, rotten crops all over Europe, and famine, riots, and epidemics that caused hundreds of thousands of deaths. When "incessant rain often confined us for days to the house," as Mary Shelley remembered, Byron proposed a competition for who could write the best horror story, and eighteen-year-old Mary won.[109] Byron never returned to England, and he didn't see his daughter again. He died seven years later after he had sailed to Greece to support the Greek War of Independence. Byron's daughter, Ada Lovelace, inherited his creativity, though, and became the world's first computer programmer in 1843, when she wrote an algorithm to be implemented in Charles Babbage's mechanical Analytical Engine. As this digression illustrates nicely, in the years after Schelling, Romanticism and science became closely intertwined, giving rise to an epoch accurately described by Richard Holmes as "Romantic Science" or an "Age of Wonder," characterized by "the notion of an infinite, mysterious nature waiting to be discovered or seduced into revealing all her secrets."[110]

It also was Schelling's monism that inspired the Danish physicist Hans Christian Ørsted to look for a relation between electricity and magnetism. As Ørsted realized in 1820, a compass needle could be deflected by an electric current, a discovery that eventually led to the unification of electricity and magnetism and the concept of field theory. Yet, the crucial step toward modern physics, with just the right blend of proximity to and distance from Schelling's philosophy, an ideal mix between down-to-earth empiricism and lofty theory building, occurred in England.

After he had returned from Germany, the poet Samuel Coleridge delivered lectures on German Romanticism and Schelling's philosophy at the Royal Institution, an organization for teaching science to laypeople. Another lecturer at the Royal Institution was his close friend—second only to Wordsworth—the chemist Humphry Davy. Davy was a nature lover and pioneer of electrochemistry who wrote pantheistic poems about how "nothing is lost" in nature and how "all the system is divine."[111] The first to isolate and discover several elements with electrolysis and to find out about the pain-relieving properties of laughing gas, he later became president of the Royal Society. Coleridge claimed that Davy had adopted his suggestion that "all composition consisted in the balance of energies," which Coleridge himself had adopted from Schelling.[112] Davy's greatest discovery, though, as he believed himself, was his assistant, a certain Michael Faraday, whom he first met when the young man attended his lectures at the Royal Institution.

Indirectly, via Ørsted and Coleridge, Schelling thus also influenced Faraday's efforts to conversely convert magnetism into electricity as well as Faraday's idea of describing electromagnetism as a field. The latter would become a crucial step in the development of quantum mechanics and finally quantum field theory. Having learned about Faraday's breakthrough, Schelling even envisioned that chemistry would also finally be reduced to electromagnetism. Ørsted's and Faraday's discoveries marked significant steps toward the unification of electricity and magnetism in James Clerk Maxwell's theory of electrodynamics, finally evolving into Einstein's theory of relativity. Being self-taught and knowing only a little math, Faraday couldn't summarize his intuitions in mathematic equations. This was eventually achieved by Maxwell in his famous set of equations.

At least equally strong was Schelling's impact on biology and the sciences of complex systems: Schelling's credo—"to grasp nature in her unity," to understand nature as an "organism" rather than a "machine"—resonated deeply with researchers such as the polymath, naturalist, geographer, and explorer Alexander von Humboldt, who credited Schelling with having stimulated a "revolution" in science.[113] Following Schelling, Humboldt understood nature as a "net-like intricate fabric," a paradigm that inspired him when he discovered that volcanoes have subterranean connections and shaped Earth's surface and that plants and animals form ecosystems and affect, for example, the climate.

Schelling's idea of opposing polarities also influenced Charles Darwin's theory of biological evolution, for which both Darwin and Ernst Haeckel would credit Schelling's influence. Already Darwin's grandfather Erasmus Darwin's 1803 poem "Temple of Nature"—full of Platonic motifs and early evolutionary thinking—described how "Immortal Love! Who ere the morn of Time, . . . gave young Nature to admiring Light! . . . Press drop to drop, to atom atom bind, Link sex to sex, or rivet mind to mind."[114] Haeckel, inspired by Humboldt and Darwin to conceive of nature as "one unified whole," was a researcher who explored nature from mountain peaks to the deep sea.[115] Unsure for some time whether he should pursue a career as an artist or as a scientist (to the dismay of his parents who wanted him to become a medical doctor), he collected flowers, moss, algae, jellyfish, and, in particular, single-celled marine microorganisms known as "radiolarians," which he documented in beautiful drawings and watercolor paintings. He characterized these objects as "delicate works of art" and "sea wonders," and they would later lend inspiration to the emergence of art nouveau.[116] Haeckel also generalized Darwin's evolution to explain the origin of humanity, reasoned that there must be intermediate species between ape and human, and inspired his students to discover the remains of Java Man, one of the first hominids ever found. One of the most important implications of genetics, Haeckel argued, would be that both animate and inanimate phenomena are part of the same nature, entailing a "unity of nature."[117] Building on the insight that the various forms of energy can be converted into each other while the total energy remains conserved, in his book

The Riddle of the Universe, he deduced a "unity of all natural forces" that has become "a guiding star leading our monistic philosophy through the vast labyrinth of riddles of the universe to its final resolution."[118] Following Spinoza, Haeckel saw "in the universe only a single substance that is God and Nature at the same time" and encouraged scholars and artists to "glorify nature or the universe" as a "vast, omnipotent wonder."[119] Haeckel developed into a popular champion of monism, and in 1904 an international congregation of freethinkers in Rome announced him as a monistic "antipope" before the group marched to Giordano Bruno's monument at the Campo de' Fiori to deposit a laurel wreath.

A Blessing and a Curse

But, of course, Schelling, Hegel, Hölderlin, and Goethe didn't know about entanglement and decoherence. They thus resorted to metaphors in order to illustrate what they meant: nature is understood as a "living organism," and evolution is understood as "creativity." The universe is equated with "God," and the unification of polarities is compared with "love." These metaphors paint a beautiful, poetic image of nature and provide intuitive and sometimes illuminating illustrations that can inspire new ideas. But if taken too literally, they obviously lead astray. This is even more problematic if such romanticizing gets combined with a contempt of experimentation, as sometimes shines through in the writings of Goethe, Schelling, and other Romanticists, a sentiment that identified experimental exploration with violence and vivisection. Thus, when Schelling's Romantic "natural philosophy" became tremendously influential, for the developing modern sciences it meant both curse and blessing.

According to Massimi, "Schelling, and 'Naturphilosophie' in general, have often been accused of being too speculative, too obscure, too alien to the experimental method to ever be in the position of exercising any genuine influence on the sciences of the time."[120] The German physicist Paul Erman, for example, described Schelling's philosophy as "deceptions and lies," and the chemist Justus von Liebig compared it to "the pestilence, the black death of the century."[121] As historian of science Richard Holmes puts it, it "constantly teetered on the brink of idiocy."[122]

From a modern perspective, there seems to be good reason to agree with Erman and Liebig. Today, the concept of an all-encompassing unity appears as half forgotten, esoteric, and even strangely totalitarian. When Ernst Haeckel, for example, naively applied Darwin's theory of evolution to humanity, he became a vanguard for social Darwinism and biological racism, which later was zealously integrated into the biologistic, brutally racist, and eventually homicidal ideology of the Nazi movement. Also the glorification of nature was eagerly hogged by Nazi ideologists. Norwegian Nobel laureate in literature Knut Hamsun's novel *Pan*, for example, full of pantheistic reflections on nature, got adapted as a movie by Nazi chief propagandist Joseph Goebbels, who considered Hamsun one of his favorite authors.

Schelling's friend and collaborator Hegel later applied their common idea about the reconciliation of polarities to history and politics. Interpreted from a materialistic perspective by Hegel's student Karl Marx, this concept became a cornerstone of the theory of Marxism. Thus the ideological background of the twentieth century's most devastating totalitarian movements may also be associated with Schelling's philosophy. More generally, the credo of the Romanticists, who stressed the priority of the creative subject over objective facts, seems to favor a development fostering alternative facts and pseudoscience—from astrology to homeopathy to antivaccination beliefs and climate change denial. Where "the One" is still discussed today, it soon gets dismissed as "new age bullshit" or hijacked by esotericism. It seems not to fit into the successful paradigm of analytic philosophy.

While such negative implications can't be denied, on the one hand, the backbone of such pseudoscientific and totalitarian ideologies is a view of nature as it—in the eyes of their proponents—*should* be, not as it actually *is*. Such mind-sets are based on imposing a predetermined image of "the One" onto the world, instead of undertaking the attempt to honestly reconstruct "the One" out of its appearance as a dazzling array of multiple things. While for Heinrich Heine "the One" was still a concept of reason, as he affirmed when he dismissed the "pious, smug-hearted Jacobi" as a "mole," the concept, seemingly discredited by its association with the horrible crimes and catastrophes initiated by the autocratic regimes of the twentieth century, sank into oblivion.[123]

* * *

What remained of the monistic paradigm is the embrace of nature and the quest for unification and beauty. Throughout the development of modern science, these threads have influenced this process just as much as the empirical method. It was the Renaissance philosophy of da Vinci and Ficino that delivered the narratives and questions motivating the Scientific Revolution to unfold in the Age of Enlightenment. Likewise, it was the nineteenth-century Romanticism of Goethe and Coleridge and "Romantic Science" such as electrodynamics and thermodynamics that set the stage for relativity and quantum physics, the great scientific revolutions of the twentieth century. Fact is, unification and the quest for mathematical beauty have become an integral part of modern physics. As Nobel Prize winner Frank Wilczek declares most emphatically in his book *A Beautiful Question*, "The only fitting answer to [the] Question 'Does the world embody beautiful ideas?' . . . is a resounding Yes!"[124] But are these notions about mathematical beauty and unification part of nature or simply wishful thinking? Are they personal preferences projected by scientists onto their work—a simple consequence of our brain being hardwired to identify patterns and symmetries that delude us into believing in a fake reality? Do they lead us astray from what is really going on in nature?

While not all conceptions inspired by a monistic harmony of the universe turned out to be correct—a prominent example being Kepler's music of the spheres or Newton's speculations about alchemy—Kepler and Newton would not have made their groundbreaking discoveries if they hadn't been striving for a harmonious, unified description of nature. In fact, according to history, whenever monism flourished, the arts and sciences thrived. In hindsight, this is not surprising: after all, more often than not creativity boils down to the discovery of unknown connections and similarities between what hitherto have been thought of as separate realms. A mind-set that adopts nature as a unity is particularly prone to find and exploit such correlations.

If, on the other hand, monism eventually follows from quantum mechanics, how is it related to the search for beautiful patterns in nature? This question directs us to the problem of identifying the fundamental description of nature and how its properties get imprinted on less fundamental layers of description. As we will see, these signatures of a fundamental reality may appear as patterns and symmetries. That doesn't imply of course that every aesthetic ideal that has been conceived is realized in nature, but it provides a justification for searching for such patterns. The three-thousand-year-old concept of monism may actually help modern physicists in their struggle to make sense out of black holes, the Higgs boson, and the early universe.

6

ONE TO THE RESCUE

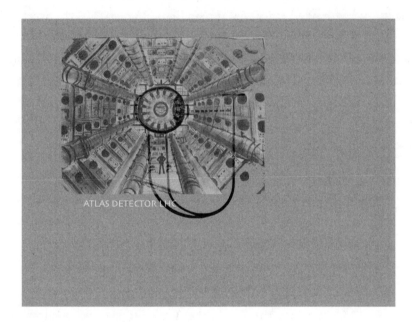

ATLAS DETECTOR LHC

W E ARE READY NOW TO ARRIVE AT THE KEY POINT of our inquiry. Given that monism follows from quantum mechanics taken seriously as a theory about nature, having understood how the plurality of things in our daily-life experience emerges, and grasping why this implication has been ignored for so long even as it was etched into the practice of science in the past, we can ask now what these insights entail for the problems and challenges physics faces today and in the future. We are prepared to uncover a new foundation of science.

Physics in Trouble

On October 18, 2017, Gian Giudice announced that particle physics was in trouble. "There are many indications that, following the recursive pattern of scientific revolutions, we are now witnessing the beginning of the phase of crisis... This is the most complex and intense moment of scientific research, when revolutionary and unprejudiced ideas are needed for a real paradigm change."[1] In the world of physics, Giudice's opinion carries more weight than most. He serves as the head of the Theoretical Physics department at CERN, the European center for particle physics in Geneva that is operating the Large Hadron Collider (LHC). As such, CERN spearheads the global effort to reconstruct the entire universe from simple, fundamental building blocks, preferably as simple as possible. And, for particle physicists, simpler always means smaller.

The great accelerators constructed by particle physicists have been compared to the magnificent cathedrals of the Middle Ages. This is a reference not just to their magnitude but to their role as a nearly sacred space for seeking if not God then at least the foundational reality and the harmonious laws that govern the cosmos. Among these, CERN's LHC is the greatest accelerator in history: twenty-seven kilometers long, crossing the border between France and Switzerland four times, it is the largest machine ever built by humankind, constructed to make the Standard Model, which summarizes the present knowledge about particle physics, complete and to discover what lies beyond. The LHC was expected to explain why the best theory in the entirety of science features awkward coincidences that make it look like a pencil balancing on its tip. And indeed, in 2012 the Higgs particle was found, the last missing piece in the Standard Model needed to provide masses both to itself and to the other massive components of nature. After that, though, the enthusiasm slowly faded. Once the Standard Model was completed, no sign of new physics to signal the reason for its shortcomings came into sight. After more than a decade of running, the LHC is still doing what it did when it was switched on: it confirms the Standard Model, over and over again, but it finds nothing else or beyond.

The problem with this situation is that there is a desperate need for new physics beyond the Standard Model. According to what astrophysicists and cosmologists have found, roughly 85 percent of the universe's matter content is made of nonluminous, dark matter that's observed only via its gravitational pull. Yet there exists no particle candidate that could chip in to explain this missing mass. What's more, matter itself only provides less than a third of the cosmic energy budget. The major part of the universe's energy is a mysterious dark energy assigned to the vacuum (i.e., empty space), driving the universe apart and into an accelerated expansion. Dark energy is even more difficult to explain with the Standard Model or simple extensions than dark matter. And then there are the coincidences: Why do ordinary matter, dark matter, and dark energy all contribute comparable amounts (within a factor of ten or twenty) to the universe's energy budget? Why is the strong force, as far as we know, equally strong for particles and antiparticles of different spin orientations? Why is the mass of the Higgs particle so small? And why is the amount of dark energy in the universe so tiny?

These observations, and most importantly the last two among them, have given particle theorists a headache. Simple estimations of the Higgs mass—including its contribution from vacuum fluctuations (i.e., virtual particles that pop in and out of existence in the vacuum)—would have it seventeen orders of magnitude too big, while the expected dark energy of the universe, estimated from the known sources of vacuum energy, would be some mind-boggling 120 orders of magnitude too big—a one followed by 120 zeroes. Obviously the Standard Model, as successful as it is elsewhere, is missing a crucial part of reality.

When physicists built the LHC, they believed that new particles would be the solution. Supersymmetry (SUSY), for example, would easily double the particle content of the Standard Model and could have provided both an attractive dark matter candidate and a solution to why the Higgs is so small. Supersymmetry postulates a symmetry between matter and forces; thus, for each known matter particle (a "fermion," as physicists say, such as quarks and leptons), there would be a yet-undiscovered force carrier particle (a "boson" in physics lingo), and vice versa. Since both types of particles contribute to vacuum fluctuations of the Higgs with opposite signs, the contributions may simply cancel out. A similar mechanism may be in operation to keep the dark energy of the universe small, although that doesn't work quite as well. SUSY, physicists realized, may contribute a major step in making the Standard Model look natural again. Ever since, thus, an increasing number of particle physicists have expected to find the missing SUSY partners—an expectation that simply has not been realized. This situation has sparked a staggering crisis in fundamental physics and has driven CERN's chief theorist, Gian Giudice, to urge his colleagues to seek alternatives: "A new paradigm change seems to be necessary," Giudice writes. "We are confronted with the need to reconsider the guiding principles that have been used for decades to address the most fundamental questions about the physical world."[2] So what are these guiding principles that need to be reconsidered? What are the beacons particle physicists hope will help them to make sense of the world?

Multiverse Versus Uglyverse

"The most incomprehensible thing about the universe is that it is comprehensible," Albert Einstein once said.[3] It probably is the most obvious symptom of the crisis particle physics endures that recently it is far from being a matter of consensus anymore that the universe is indeed comprehensible. Can the universe still be considered simple and elegant, is this apparent elegance just accidental instead of fundamental, or is the universe even messy and our perception of beauty just based on prejudice? Whereas, in the past, particle physicists were guided by the expectation of living within a unique universe, governed by simple laws and symmetries, this prospect is increasingly disputed now. Up for debate are two popular yet highly controversial concepts challenging this view, characterized tellingly by the buzzwords "multiverse" and "uglyverse."

Multiverse proponents advocate the idea that there may exist innumerable other universes, some of them with distinct particles and forces and even different numbers of spatial dimensions. This multiverse refers not to the alternative realities of Hugh Everett's interpretation but to other patches of space-time governed by entirely different laws of physics, coming out of string theory, which seems to allow for a vast "landscape" of possible universes. If now inflation, the phase of accelerated cosmic expansion that most physicists believe gave birth to the universe, goes on long enough (or maybe forever, as many models predict), it is producing a large (or even infinite) number of bubble or baby universes, and among them, all the potential universes of string theory could become real. According to this rationale, if the coincidences mentioned above seem to favor the emergence of complex structures, life, or consciousness, we shouldn't be surprised to find ourselves in a universe that allows us to exist in the first place. Just like fish have no reason to wonder why they live in water, the fact that we observe a small Higgs mass and a tiny dark energy may be a prerequisite of our own existence. That would suggest we don't have to worry about weird coincidences in our universe anymore. This "anthropic argument" implies that we find ourselves in a life-friendly spot since otherwise we wouldn't exist in the first place and there would be no one to observe this inhospitable universe.

What, then, is the problem with the string theory landscape, if we have already embraced Everett's many worlds? After all, in Everett's many worlds interpretation, quantum mechanics spawns a multiverse quite naturally. Firing individual electrons at a screen with two slits results in an interference pattern on a detector behind the screen. In each case, it appears that the electron goes through both slits each time. As we have discussed in Chapter 3, decoherence can convert these potential trajectories into parallel realities or Everett branches. So maybe, in similar fashion, also the string theory multiverse is just a completely natural yet unfamiliar part of physics.

Yet, while quantum mechanics according to Everett gives rise to a multiverse of alternative realities, all subjected to the same set of physical laws, string theory—claiming to be a fundamental theory of everything—seems to predict a multiverse of universes governed by different laws of physics altogether.

Thus, as natural as the assumption of a multiverse appears, "anthropic reasoning" can create serious problems, as it eventually implies that we may have a hard time predicting anything anymore. Ever since the early days of science, finding an unlikely coincidence has prompted an urge to explain, a motivation to search for the hidden reason behind it. With the advent of the string theory multiverse, this has changed. As unlikely as a coincidence may appear, in the zillions of universes that compose the multiverse, it will exist somewhere. There is no obvious guiding principle for the CERN physicists searching for new particles. And there is no fundamental law to be discovered behind the accidental properties of the universe. Thus "the multiverse may be the most dangerous idea in physics," argues South African cosmologist and famous collaborator of Stephen Hawking George Ellis.[4]

Quite different but not less dangerous is the other challenge. "Uglyverse" is a snappy label *New Scientist* magazine assigned to a proposal by theoretical physicist Sabine Hossenfelder. According to her popular book *Lost in Math*, modern physics has been led astray by its bias for "beauty," giving rise to mathematically elegant, speculative fantasies without any contact with experiment: "Why should the laws of nature care what I find beautiful? Such a connection between me and the universe seems very mystical," Hossenfelder righteously objects.[5]

Of course, nature could be complicated, messy, and incomprehensible. However, what physicists call "beauty" are structures and symmetries. If we can't rely on such concepts anymore, the difference between comprehension and a mere fit with experimental data will be blurred. If science had to abandon any reliance on simplicity and aesthetic appeal, its result could be produced by a computer finding a set of mathematical equations that provides the best description of observations but would be arbitrarily complicated and devoid of any intuitive grasp of what's going on.

Just like the multiverse, science without beauty may sacrifice our ability to make sense of the universe. By facing the challenge of whether we may still find a natural explanation for cosmic coincidences, we have arrived at a "dramatic bifurcatory moment in fundamental physics," according to Nima Arkani-Hamed of the Institute for Advanced Study in Princeton, New Jersey, one of the most influential theoretical physicists today.[6] "Is nature unnatural?" asks science writer Natalie Wolchover, who worries that "the universe might not make sense."[7]

So let's take a step back and ask, How do physicists traditionally address the most fundamental questions about the physical world? And what could have possibly gone wrong? The answers to these questions are intimately intertwined with the way we grasp the universe. And it becomes most obvious when we remember how we first started to make sense of the world around us, how we were as children, when life was a game. In many respects, physicists still are just like little kids.

Playing with Bricks

Kids are enamored with bricks. Bricks can become everything: a bridge, a castle, a house for a loving family, a beauty salon for dolls, a pirate ship or spaceship, whatever a child's effervescent fantasy comes up with. And when one deconstructs these things, the same bricks can be rearranged to become the next toy. We never really grow out of this fascination with imaginative power. And maybe that is what makes us human, after all.

How are bricks and the many things they can be arranged into related, and what does that tell us about reality? An important aspect in this context is language and the many ways we can talk about exactly the same things. In his best-selling book *Sapiens: A Brief History of Humankind*, Israeli historian Yuval Noah Harari argues, "The truly unique feature of our language is not its ability to transmit information about men and lions." Rather, Harari suggests, "it's the ability to transmit information about things that do not exist at all."[8] As Harari insists, "Only *Homo sapiens* can speak about things that don't really exist, and believe six impossible things before breakfast."[9] Following Harari, spaceships or dollhouses are just bricks charged with fantasy. And this kind of play isn't just for kids; it is a genuine part of human nature: "Law, money, gods, nations," Harari continues, "any large-scale human cooperation—whether a modern state, a medieval church, an ancient city or an archaic tribe—is rooted in common myths that exist only in people's collective imagination."[10] As Harari argues, this ability "to create an imagined reality out of words enabled large numbers of strangers to cooperate effectively" and eventually "opened a fast lane of cultural evolution, bypassing the traffic jams of genetic evolution."[11]

It is an intriguing hypothesis. But can we really deny entities such as money, stock corporations, or laws as parts of reality? After all, a lack of money, a crash of the stock market, or a violation of law can get us into serious trouble. Moreover, if such "fictions, social constructs or imagined realities" exist only in people's imaginations, we must extend the same critical spirit to people and biological organisms themselves.[12] Do they, too, really exist?

Enter Erwin Schrödinger. For the quantum pioneer who had been competing with Werner Heisenberg for the correct interpretation of quantum mechanics, Adolf Hitler's rise to power meant the beginning of an odyssey. First, despising the Nazis, Schrödinger left Germany in 1933 for a position in Oxford, but the traditional, square college town didn't accept his living together with Anny and another woman in a ménage à trois. So he decided in 1936 to move back to his native Austria, which was, however, occupied by the Nazis in 1938. As a consequence Schrödinger, now considered "politically unreliable," found himself fired. Schrödinger packed his bags again and finally settled in Dublin, Ireland, where he gave a set of public lectures pondering the question of what "life" really is.

In this effort to provide a definition of life, he described the process of living as a property of matter. "What is the characteristic feature of life? When is a piece of matter said to be alive?" Schrödinger asked and suggested, "When it goes on doing something, moving, exchanging material with its environment, and so forth, and that for a much longer period than we would expect an inanimate piece of matter to keep going under similar circumstances."[13] Schrödinger transformed his lectures later into an influential book, where his characterization of life through its striking stability helped to pave the way for the field of molecular biology. Both James Watson and Francis Crick confirmed that they were independently inspired by Schrödinger's book to decode the structure of DNA.

But most important here is Schrödinger's defining life by its functionality rather than by its material basis. For Schrödinger, a living organism had to be understood as an information-processing routine rather than an object in space and time. Indeed, Schrödinger's notion is supported by the fact that over a typical human life span, most atoms in the human body will be replaced without altering the identity of the individual.

The foundational spirit behind Schrödinger's thesis is of course not restricted to the explanation of life but can be generalized to any kind of macroscopic phenomenon. A beautiful account of this point of view can be found, for example, in the travelogues of the eighteenth-century explorer, naturalist, and revolutionary Georg Forster. When he describes in poetic language how the waves of the ocean "arise and tower up, they froth and disappear; they are engulfed again by the vastness. Nowhere is nature more terrible than here in the unrelenting severity of its laws," Forster feels and realizes that, at the same time, "nowhere does one feel more clearly that, set against the whole of matter, the wave is the only thing that passes through a point of separate existence, from nonbeing back into nonbeing; yet, the whole rolls forward into immutable unity."[14] Again, this philosophy can be traced back to Plato, who in his *Timaeus* understood the foundational elements comprising nature as informational patterns imprinted into a material basis or foundation, the "midwife of being." By comparing the *Timaeus* to Plato's *Parmenides*, it is easy to identify this midwife of being with "the

One," the hardware of the universe. In this sense waves or biological organisms are more akin to what Harari calls "social constructs" than to matter itself. So, after all, if we deny corporations are part of reality, we have to deny reality to tsunamis and ourselves as well.

Maybe it makes more sense to perceive the world in terms of many realities rather than a single, exclusive one. "There is only one world, the natural world," but "there are many ways of talking about the world . . . Our purposes in the moment determine the best way of talking," emphasizes California Institute of Technology cosmologist Sean Carroll in his book *The Big Picture*.[15] Just like little kids, depending on what we want to play, we talk about bricks as beauty salons or castles. Once we embrace the concept of many realities, on the other hand, new questions arise. Do all these realities exist on an equal footing, or is there a layer of description that is more fundamental or "prior" to other ways of description?

One World . . . Many Ways of Talking

So, what is more real, the forest or the trees? The living organism or the atoms it is made of? Around 2010, American philosopher Jonathan Schaffer was pondering these problems, and he had a revelation: these questions don't make sense!

The actual difference between these concepts isn't whether they are more or less real but whether they are more or less fundamental. "Consider a circle and a pair of its semicircles," Schaffer proposed and asked himself, "Which is prior, the whole or its parts? Are the semicircles dependent abstractions from their whole, or is the circle a derivative construction from its parts?"[16] Next, Schaffer went on and generalized his problem to the entire universe: "Now in place of the circle consider the entire cosmos (the ultimate concrete whole), and in place of the pair of semicircles consider the myriad particles (the ultimate concrete parts). Which if either is ultimately prior, the one ultimate whole or its many ultimate parts?" He arrived at an unexpected conclusion: the fundamental layer of reality is not built upon constituents but the universe itself—understood not as the sum of things making it up but rather as a single, entangled quantum state. Schaffer takes the stance of monism and explains, "The monist holds that the whole is prior to its parts, and thus views the cosmos as fundamental, with metaphysical explanation dangling downward from the One." Similar thoughts had been expressed earlier—for example, by Heisenberg's friend and student Carl Friedrich von Weizsäcker. According to Schaffer and Weizsäcker, on the most fundamental level of description, there exists only one single object: the quantum universe.

But, if this is so, doesn't it once again return us to the same old problem, that it doesn't seem to make sense to say that corporations, tsunamis, and we ourselves aren't real? As Schaffer readily admits, one might argue that "on such a view there are no particles, pebbles, planets, or any other parts to the world. There is only the One. Perhaps monism would deserve to be dismissed as obviously false, given this interpretation."[17] But Schaffer disagrees, as such a notion of monism is based on a fundamental misunderstanding: "The core tenet of historical monism is not that the whole has no parts, but rather that the whole is prior to its parts."[18] Repeatedly, we have characterized matter, space, and time as "illusions." But what do we really mean by "illusions"? Obviously we don't mean that these are a question of choice, that we simply can decide whether or not we want to live in a universe governed by space, time, and matter. In fact, when physicists sloppily refer to time "as an illusion," they usually mean it's not a fundamental property of the universe; it's a property of our perspective on the universe. That doesn't mean, of course, that time doesn't have real consequences, and one may argue that "illusion" is just not the proper word. Carlo Rovelli once wrote me that he prefers to call time "perspectival" instead. Yet, a fata morgana is a real perception that can even be photographed. So the illusion in the case of a fata morgana is that the real perception is associated with a nonexisting property of the landscape in front. In the case of time, as we will see in Chapter 7, the illusion is that our experience of time is associated with an arguably nonexisting property of the fundamental universe. It is no part of the description that is "prior," as Schaffer understands it.

So what does "prior" really mean here? For kids, the answer is simple: the bricks are your fundamental building blocks; the more complex arrangements in the shape of ships or houses are what bricks can produce. Likewise, from a particle physicist's point of view, these fundamental bricks would be quarks, electrons, and neutrinos, plus some quanta to let them interact and the Higgs to endow them with masses. This account is oversimplified, of course. In modern particle physics, particles are understood as excitations of quantum fields rather than brick-like objects. It would be quantum fields rather than particles that provide the fundamental building blocks of the universe. Still, in the prevailing paradigm, more fundamental fields are expected to crop up at higher energies. After all, this is the reason why particle physicists employ accelerators in their exploration of the foundations of physics: the foundation of physics is expected to be found at the highest energies, corresponding to the smallest-possible distance scales. According to a particle physicist's philosophy, more fundamental means simpler, and simpler means smaller.

When kids rejoice about the mutability of bricks becoming whatever they like, they usually don't wonder why it doesn't work the other way around. However you deconstruct your brick-built castle or pirate ship, you will get back your same old bricks; you won't end up with a set of Barbie dolls or Matchbox cars. For entangled quantum states, though, this is different: if you plug two particles together, one left- and one right-spinning, you can decompose the product you get not only back into the same two particles but equally well into a particle that spins around a vertical axis and another one that rotates in the opposite direction. In this case, it gets fuzzy to determine what is really fundamental, constituents or compounds. Is there a possibility of deciding this question, to tell apart what's fantasy and what is reality in this case?

Philosophers describe such relationships between more or less fundamental descriptions of nature with the term "emergence." Weak or strong emergence denotes cases where the natural laws of one level of description (such as biology or pirate ships) cannot—either in practice or fundamentally—be derived from laws on a more fundamental level (such as atomic physics or bricks). While the existence of weak emergence is unchallenged—nobody wants to describe the stock market by calculating the behavior of the elementary particles involved—accepting strong emergence, on the other hand, is not too different from believing in miracles. After all, if phenomena on a higher, more complex level are not even in principle describable on a more fundamental level, then the fundamental level does also not constrain the space of possibilities of the higher level. For example, if the fact that particles can't propagate faster than light according to Einstein's theory of relativity does not imply that also biological organisms can't run at superluminal speed, there is also nothing that prevents Jesus from walking on water. Any extrapolation of known physical laws, meaning any application of these laws in new situations, would become questionable. In this case the entire scientific endeavor wouldn't make sense anymore. In fact, it is this insight that makes perfectly clear how crucial it is for science to rely on a stable foundation.

Once we exclude the possibility of strong emergence, it becomes apparent that not all realities are equally fundamental. Rather, the more fundamental realities constrain the space of possibilities for the higher levels: while more fundamental realities or natural laws are still valid on higher levels, more complex or higher-level descriptions have a limited range of applicability when extrapolated to more fundamental constituents. Obviously the science of sociology only makes sense from a viewpoint that accepts the existence of biological organisms beforehand, while particle physics in contrast does not rely on any such premise.

This view suggests a hierarchy of sciences similar to the one postulated by the French philosopher Auguste Comte. In this hierarchy physics defines the foundation, chemistry is the physics of the outer atomic orbits, biology deals with the chemistry of complex organic molecules, psychology describes the biology of the neural system, and sociology and economics discuss the psychology of large numbers of individuals. Naively it seems natural that realities are the more fundamental the smaller their defining constituents are. Yet size is not really what matters here.

The closer we look, the more we come to realize that we have to revise our traditional concept of reductionism. Once we reduce sociology and psychology to biology, it is obvious that reductionism must be understood no longer as a materialistic approach but rather as a concept in information theory: a football team is the same team, even if a player is sold, and a nation-state is the same state, even if the foreign affairs minister gets fired. The reason is that the functioning of a football team or state doesn't rely on all traits of the individual players or politicians. If an individual, for example, prefers toffee over plain chocolate, this usually won't affect his fitness or diplomatic skills.

What is important here is that physicists—or scientists in general—are masters of describing nature in various degrees of accuracy. Whenever possible, they will consider only the information relevant for the problem they want to solve. For example, when physicists want to calculate the trajectory of a cannonball, it is sufficient in most cases to adopt the cannonball as spherical, even if there exist tiny deformations. Ignoring these tiny deformations makes the physicist's life easier and the equations that describe where the cannonball will hit solvable. In this case the microscopic physics is more fundamental than the physics of spherical cannonballs. It is also less useful. In the example of the deformed cannonballs, reductionism defines higher layers of description by their functionality (in this case to follow a predicted trajectory and hit and destroy a target). The successful description of complex objects depends on how to identify the information most relevant for their specific purpose and ignore everything else. In contrast the more fundamental layer of description doesn't discard any information.

A description of reality (or simply "a reality") is thus the more fundamental the less information it discards. Given that the neglect of information is reasonably determined by "our purposes in the moment" that "determine the best way of talking," as Carroll writes, the more fundamental reality is also less dependent on an actual objective or perspective; it is more observer independent. Conversely, higher layers of description are typically built on concepts that are more substrate independent, relying more on the processing of information than on the properties of their actual constituents, and in this sense they are also more idealized.

A Quantifier of Ignorance

Physicists even have developed a measure for the amount of information they are neglecting, a quantity called "entropy." Entropy is often misunderstood as a quantifier of messiness. This has a reason: while an unbroken cup is in a state with low entropy, the broken pieces of a cup that has fallen down from the table are in a state of high entropy. Yet the correct definition of entropy is given by the number of "microstates" corresponding to a "macrostate." Now what does that mean? Simply put, a "microstate" is a state that is completely known and specified; a macrostate is a state that isn't.

Take, for example, a car. A car can be either intact or broken. But while there is one possible state in which the car is intact (all parts in the place they are supposed to be), there are many ways in which a car can be broken (a wheel missing, a shattered windshield, a battered chassis, etc.). In this example, the macrostate "intact" has a low entropy, while the macrostate "broken" has a high one. This explains the relation with messiness: in a tidy room we know where everything is (low entropy)—namely, exactly where it belongs. In a messy room, though, we don't know where things are; we just know they are not where they belong. Thus a messy room features a high entropy since the information where things are is lost.

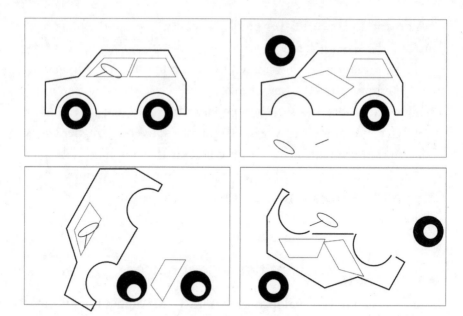

There are many ways for a car to be broken but only one possible state in which the car is intact. The macrostate "broken" thus corresponds to many microstates, implying a large entropy, while the macrostate "intact" corresponds to a single microstate only, having a vanishing entropy.

To illuminate this point, we can take up the prototypical example of an emergent theory: thermodynamics in its incarnation as statistical mechanics. In thermodynamics, states such as gases or liquids are described by parameters such as temperature, pressure, or volume. As common as these quantities are, they are not fundamental in the sense that they do not correspond to a specific configuration of the constituent atoms or molecules (i.e., the microstate). Instead, they characterize a statistical average of microstates, the macrostate. The temperature of a gas, for example, can be related to the average energy of the constituent atoms.

The concept of entropy describes how many concrete microstates correspond to a given macrostate.[19] If we have a pot full of water heated to two hundred degrees Fahrenheit, its entropy tells us how many different configurations exist for the water molecules in which they have exactly specified energies that amount to the same average of two hundred degrees. Thus entropy can be understood as the missing information to identify the exact microstate in a given macrostate. Now remember that we defined a fundamental description of reality by its feature that no information gets discarded. The concept of entropy thus allows us to find a concept of fundamentality that is more general than the spoiled constituent concept of traditional reductionism. A microstate is fundamental; a macrostate is not.

Consequently, the fundamental state of the universe has zero entropy. Turning back to quantum mechanics, it is well known that when we identify a component in an entangled quantum system, the so-called "von Neumann" entropy increases as a consequence of the information loss that corresponds to the now ignored link of the component with its environment. Identifying components implies discarding what unifies the universe into a whole, and discarding information means we are adopting a coarse-grained rather than a fundamental perspective. As a consequence, the fundamental state of the universe cannot be a constituent; it has to be the total entangled system including everything—observer, measured system, and environment—also known as the quantum universe itself. It is this concept, that the foundation of science should be based on the most fundamental description, that implies the relevance of monism for physics. It also implies that the traditional approach of looking for the foundations of nature at high energies or by identifying increasingly small constituents is misguided.

This flaw is becoming obvious even in string theory, the most radical approach to explaining the universe in terms of tiny building blocks. As string theory pioneer Leonard Susskind points out, in "String Theory ... nobody knows ... what the basic 'building blocks' are."[20] Depending on the specific solution of the theory studied, sometimes the eponymous one-dimensional strings appear as fundamental, while at other times higher-dimensional objects known as "branes" do so. Susskind describes "the landscape," the realm of possible solutions to the theory, as "a dreamscape in which, as we move about, bricks and

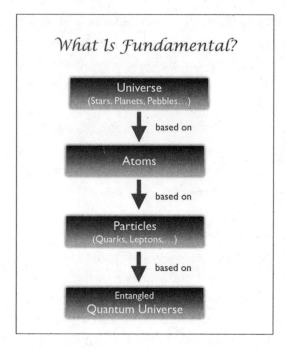

The hierarchy of fundamentality: the many things populating the universe are made of atoms, which again are made of more fundamental particles. Particles themselves, however, are abstractions from the most fundamental quantum universe.

houses gradually exchange their role. Everything is fundamental, and nothing is fundamental . . . The answer depends on the region of the landscape we are momentarily interested in."[21]

This finding becomes highly problematic when we recall why particle physics or string theory is pursued in the first place. Particle physics is a many-billion-dollar endeavor with few practical benefits. It is justified only if it defines or at least contributes to exploring the foundations of science. If these foundations are shaky, so is the excuse for investing money and brain power into these fields. But the fact that string theorists are failing to identify a fundamental constituent within string theory doesn't imply that we have to give up the concept of fundamentality altogether. It means that by zooming in to increasingly small distances and high energies, we have lost sight of the whole—that by splitting the world apart, we have discarded the links that hold the universe together.

In fact, particles or strings are quanta, so the way we understand quantum physics determines directly how we understand particles and strings. If particle physics and string theory claim to identify the foundations of physics, they cannot rely on the Copenhagen interpretation that denies quanta are part of reality. Evidently, if quantum physics is not more than a recipe to produce predictions for classical objects, the same is true for particle physics and string theory. Such a philosophy hopelessly fails to ground the entire discipline of physics, not to mention science, in a solid foundation. If, on the other hand, quantum mechanics is taken seriously as a theory about nature, the phenomenon of entanglement implies that subsystems can't be fundamental. What is fundamental is the entangled quantum universe.

To be of any use at all, particle physics and string theory then have to be reinterpreted to be consistent with this simple consequence of quantum mechanics. Both particles and strings must be based on an even more fundamental level: the quantum universe. Admittedly, such a reinterpretation won't be an easy task. It will be necessary to study in detail how particles and strings are related to and emerge from a quantum description of the entire universe. But this effort will be necessary to allow particle physics and string theory to do what they are good for (i.e., to define the foundation of physics), and it may help to solve the problems and inconsistencies particle physics and particle cosmology are facing today.

"Natural Nature" and the Quantum Universe

So how can a monistic foundation of science be of any help with the challenges encountered in fundamental physics today? As I will argue, the troublesome coincidences that constitute the existential crisis of particle physics and cosmology today may be resolved when they are considered from the perspective of focusing on the whole instead of the part.

We start by uncovering the forgotten origin of the debated concepts of symmetry and mathematical beauty. As history has shown us over and over again, the quest for unification and mathematical beauty that gave rise to modern physics is a heritage of monistic philosophy. The victory march of unification didn't stop with Isaac Newton and James Clerk Maxwell. After it was realized in the nineteenth century that electricity, magnetism, and optics were all different aspects of the phenomenon of electromagnetism, in the twentieth century electromagnetism was unified with the newly discovered weak force that is responsible for nuclear beta decay. The strong force responsible for confining the quarks within neutrons and protons and for keeping the neutrons and protons together inside atomic nuclei was described with a generalization of the theory that describes the quantum version of electrodynamics. Finally grand unified theories that were first developed in the 1970s but haven't been confirmed in experiment so far strive to describe quarks and leptons (i.e., the various building blocks of nature) as different states of a single type of particle and the known forces, with the exception of gravity, as different aspects of a single force. The central element in this program is symmetry, and symmetry is usually invoked when scientists enthuse over the beauty of physics. According to Emmy Noether, Grete Hermann's genius supervisor in Göttingen, whenever a theory respects a certain symmetry, meaning whenever the physics remains unchanged while the elements of the theory are shifted or rotated, either in normal space or in abstract mathematical spaces, this implies the existence of a conserved quantity. Momentum, for example, is conserved since the same physics applies at different locations in space (i.e., it is symmetric under spatial shifts). An intuitive example is given by a body moving with constant speed along a flat plane that can be shifted in any direction without making a difference. A plane with slopes, dents, or speed bumps, though, on which bodies are accelerated or decelerated, is altered when it is shifted, and as a consequence momentum conservation doesn't hold anymore. Likewise, energy, for example, is conserved since the same physics holds at different times (i.e., it is symmetric under time shifts). In 1932, Heisenberg had generalized these shifts in space and time to rotations among particle species in Hilbert space. By realizing that protons and neutrons behave almost interchangeably when one is ignoring their different electric charges, he understood the two types of particles as different states of a single object. This logic is taken to the extreme in grand unified theories, where the physics remains the same when all (or a major part) of the known types of particles are exchanged with each other. In a similar spirit, supersymmetry, which became popular among particle physicists in the 1980s, aims to unify forces and matter, and string theory tries to unify all known forces in a "theory of everything," including gravity. If we are skeptical about how this approach is justified and puzzled about why it has been so successful in the past, it may be since we have forgotten where this paradigm originally came from.

It is important to realize that the quest for unification in modern physics is only a halfhearted echo of monism: the profound puzzle of why the same laws of physics seem to govern all parts and the entire history of the universe is usually explained by the assertion that everything in the cosmos is made up of the same set of particles, known as quarks and leptons. Taken at face value and subjected to critical scrutiny, this is no explanation at all. After all, nothing explains why the particles themselves behave the same way at different locations and times in the universe. Quantum field theory understands these particles not as individual objects but as excitations of quantum fields permeating the universe, which is both a better explanation and also a big step in the direction of monism in that it implies that particles are not independent of each other but all part of a unifying field. Indeed, different copies of the same quark or lepton are literally indistinguishable. Popular concepts in modern particle physics, such as grand unification and supersymmetry, go another step further by speculating that the different kinds of quantum fields encountered in nature are actually only different states of a single quantum field. Still, in contrast to Plato's One, this unified field is understood to evolve in space and time. Only quite recently it was found that the importance of entanglement in quantum field theory hasn't been sufficiently appreciated so far and that this may have far-reaching consequences for our understanding of what space and time really are.

To me, it doesn't seem unreasonable to assume that the success of guiding principles, such as unification and mathematical beauty, that are not based on experimental investigation may be grounded in the monistic implications of quantum mechanics. This brings us back to the prominent dilemmas of a comprehensible, unique universe and the desperate proposals for alternatives, the multiverse and the uglyverse. Speaking of the uglyverse, it is true that nature could be complicated, messy, and incomprehensible—if it were classical. But nature isn't. Nature is quantum mechanical. And while classical physics is the science of our daily life where objects are separable, individual things, quantum mechanics is different. The condition of your car, for example, isn't related to the color of your wife's dress. In quantum mechanics though, things that have once been in causal contact remain correlated, by fiat of entanglement. Such correlations constitute structure, and structure is beauty. Thus what is beautiful in physics may offer us a glimpse from the frog perspective onto the hidden and unique "One."

But what about the multiverse? Do we—by appealing to quantum mechanics and embracing a multiverse of many Everett worlds in order to justify the beauty of physics—sacrifice the uniqueness of the universe? As pointed out already in Chapter 3, this isn't the case. Taking quantum mechanics seriously predicts a unique, single quantum reality underlying the multiverse, as Everett and H. Dieter Zeh have already shown. Interesting in this context is that this conclusion extends to other multiverse concepts such as different laws of physics in the various "valleys" of the "string theory landscape" or other "baby universes" popping up in eternal cosmological inflation. Whatever multiverse you have, when you adopt quantum monism, they are all part of an integrated whole. As a consequence, there always exists a more fundamental layer of reality underlying the many universes within the multiverse, and that layer is unique.

As we have seen, both monism and Everett's many worlds are predictions of quantum mechanics taken seriously. What distinguishes these views is only the perspective. What looks like "many worlds" from the local observer's frog perspective is indeed a single, unique universe from the global bird perspective (such as that of someone who would be able to look from outside onto the entire universe). This insight gets reflected in recent efforts to rethink the multiverse. In 2010, Anthony Aguirre and Max Tegmark authored an article that explored quantum mechanics in an infinite universe. They concluded that in an infinite space, all potential possibilities are realized anyway, so that such a scenario provides "a natural context for a statistical interpretation of quantum mechanics" that pops out quantum mechanical probabilities automatically, while rendering a "cosmological interpretation of quantum theory" that unifies "the many worlds of Everett's interpretation . . . into one."[22] One year later, Yasunori Nomura and, independently, Raphael Bousso with Leonard Susskind developed these ideas further and concluded that "the eternally inflating multiverse and many worlds in quantum mechanics are the same," as Nomura writes,[23] or that "the many-worlds of quantum mechanics and the many worlds of the multiverse are the same thing," according to Bousso and Susskind.[24] In the context of quantum mechanics, the various concepts of multiverses are easily integrated into a unifying One.

If the fundamental description of the universe is the universe itself, perceived as "the One," this implies that science has to be founded on quantum cosmology. Physics should start with a quantum mechanical state vector of the universe, and space, time, and the Standard Model of particle physics should be derived from this fundamental description through decoherence. When I proposed this approach to Erich Joos, the decoherence pioneer and former student of Zeh, he agreed reluctantly: "Yes, in principle this is the program. The only question is what else has to be put in. That notion that the Standard Model can be 'derived' I would judge as extremely optimistic."[25] It is highly interesting in this context that leading researchers in quantum gravity have started to scrutinize the deeper meaning and the foundational role of symmetries, usually considered to be the backbone of particle physics. Edward Witten, a leading string theorist and often considered the greatest theoretical physicist in the world, finds that "in a modern understanding of particle physics, global symmetries [the reason for conserved quantities such as momentum, energy, or charge] are approximate and gauge symmetries [the principle behind forces] may be emergent."[26]

But what could gauge symmetries emerge from? Carlo Rovelli, a champion of the alternative approach to string theory, loop quantum-gravity, and a best-selling author, argues that gauge symmetries constitute "handles through which systems couple" and reflect the "relational structure of physical quantities."[27] In fact, gauge symmetries relate the existence of forces to the freedom to redefine physics differently, in different patches of space-time, and thus reveal the coherence of space-time. Building upon Rovelli and others, Henrique Gomes, a theoretical physicist and philosopher at the University of Cambridge, identifies "holism as the empirical significance of symmetries."[28] Such considerations may eventually turn out to be first efforts to link foundational symmetry patterns to a monistic quantum universe and eventually derive them from there. Meanwhile, a growing number of physicists are beginning to explore physics starting from a fundamental quantum perspective: this program is called "Physics from Scratch" by Max Tegmark,[29] "All Is Psi" by the Russian-Israeli quantum physicist Lev Vaidman,[30] and "Mad-Dog Everettianism" by Ashmeet Singh and Sean Carroll.[31] Zeh himself, in an e-mail written on April 13, 2018, only two days before he died unexpectedly, confirmed that such an approach advocated by myself had his "fullest appreciation" and "surely includes some entirely new philosophical aspects."[32]

Eventually, can a monism-based approach also resolve the fine-tuning problems encountered in particle physics and cosmology? I believe this may be possible. Remember that in physics, monism manifests itself as entanglement. Entanglement is the reason why in quantum physics, the whole is more than its parts. Just like the individual spins in David Bohm's version of the Einstein-Podolsky-Rosen paradox, which are pointing always in opposite directions since they are part of a common, entangled, spinless state, entanglement creates correlations that appear as a miracle, a ridiculously unlikely coincidence, unless the related subsystems are understood in the context of the whole. If one were just looking at these components without knowing about the total, entangled state, one might be just as perplexed about this anticorrelation as particle physicists are about the fine-tuned cancellation among the contributions to the Higgs mass.

From this point of view, coincidences such as the fine-tuned Higgs mass or dark energy of the universe shouldn't surprise us; they should be expected. The homogeneity and the tiny temperature fluctuations of the cosmic microwave background, for example, which indicate that our observable universe can be traced back to a single quantum state, usually identified with the quantum field that fuels primordial inflation, actually support this view. Based on this one can speculate how entanglement can help, for example, to understand the perplexingly small value of the Higgs mass. We originally encountered entanglement as a phenomenon occurring in quantum systems composed of many particles. But in contrast to Bohm's spinning components, in the case of the Higgs boson we are dealing with a single particle. Can entanglement play any role in this context? Is there a notion of entanglement for a single particle?

Indeed there is, as was pointed out around 2005 by various researchers in quantum information science, among them Steven J. van Enk and Vlatko Vedral, using the example of a beam splitter. A beam splitter is a technical device that divides an incoming light ray into two components and sends them off in different directions. In the case of a light ray containing only a single photon, this photon is dispatched to two different locations, a situation usually described as a quantum superposition. But as van Enk, Vedral, and others have pointed out, this situation is mathematically analogous to entanglement (it can be understood as a superposition of a "composite" state composed of the situations "particle here" and "not there" and of "particle there" and "not here"). What's more, this state can be used to prepare entangled many-particle systems, giving rise to entanglement in its original sense: "One-particle entanglement is as good as two-particle entanglement," Vedral and his coauthors conclude.[33] In Bohm's example of entangled spins, it is the symmetry or antisymmetry of the total state (its property of having a defined, total spin) that explains the correlations that are entirely surprising at the constituent level. If the apparently fine-tuned vacuum fluctuations contributing to the Higgs mass could be understood as entangled, could something similar be at work? Could there exist a hidden symmetry that would reveal itself only after all contributions of the Higgs mass have been summed up and not in the individual contributions?

When I watched Nima Arkani-Hamed, one of the masterminds of the younger generation of particle theorists, contemplating exactly such hidden symmetries to explain the fine-tuned Higgs mass while giving an online seminar in April 2021 during the COVID-19 pandemic, I asked him whether entanglement could play a crucial role in this context. "The words you say sound like they could be potentially interesting, but I think the difficulty will be that entanglement is far too generic a feature of quantum mechanics, and there are lots of other situations . . . where entanglement seems to play just as much of a (ubiquitous) role . . . , but where naturalness works perfectly," Arkani-Hamed replied, adding, "I think if such ideas are relevant, they need to take some special advantage of the Higgs . . . situation—perhaps importantly including gravity—but then it goes beyond 'just' statements about entanglement."[34]

While the final answer to my question is still outstanding, an increasing number of physicists are now starting to explore the possibility that physics at high and low energies somehow conspires in an unexpected way to produce the small Higgs mass observed. "UV/IR mixing" is the buzzword describing such theories, which are essentially, as Natalie Wolchover describes, "reexamining a longstanding assumption: that big stuff consists of smaller stuff."[35]

In fact, the problem of the tininess of the Higgs mass is often discussed in a more technical way as the problem that low-energy physics or large-scale objects should be independent of the details of high-energy physics or small-scale constituents. As Cliff Burgess, a particle theorist at the Perimeter Institute in Ontario, Canada, describes, "We can understand each of these on its own terms, and need not understand all scales at once. This is possible because of a basic fact of Nature: most of the details of small distance physics are irrelevant for the description of longer-distance phenomena."[36] This notion would imply that the exact details of the high-energy quantum fluctuations contributing to the Higgs mass would at some point become irrelevant, that they could be ignored. But can they, really?

A quantum field can be pictured as an array of quantum mechanical springs, similar to a spring mattress. These springs oscillate more or less vigorously, depending on how much energy they store. In principle, one should expect now that, just like particles at different locations can be entangled, also field or mattress states at different energies, corresponding to high- and low-energy oscillation modes, should be entangled. Following this logic, one may wonder whether entanglement between high- and low-energy field modes may be able to infuse a kind of nonseparability between high and low energy or small and large objects into quantum field theory. As has been argued recently, "renormalization," a technique to get rid of infinities popping up in calculations of finite observable quantities, may be sufficient to solve also this problem and disentangle high- and low-energy modes consistently.[37] Renormalization is a kind of trick sweeping the problem with infinities under the rug by assuming that they were included in the finite values measured all along; yet the procedure works surprisingly well. It is, however, unclear whether such kinds of arguments remain valid at energies large enough that gravity becomes relevant. Maybe at this point the entire paradigm of how to deal with quantum fields at different energies needs to be reconsidered. Arkani-Hamed, for example, likes to stress in his talks that at some point higher energies cease to probe smaller distances. The reason is that the energies involved get large enough to have the particles collapse into a black hole whose size grows with increasing energy. So, starting from this point, smaller distances are inaccessible even with higher energies; they are hidden inside the black hole, and the traditional wisdom identifying smaller distances with higher energies needs to be reconsidered.

* * *

Eventually, by arguing that the foundation of physics should be based not on particles or strings but on the monistic whole (i.e., on the entire quantum universe), and by speculating how this approach may both resolve the apparent unnaturalness and troubling coincidences particle physicists encounter in their theories based on constituents and explain the success such theories based on mathematical beauty and symmetry have had so far, we arrived at the problem of quantum gravity. Indeed, how Newton's gravity can be reconciled with quantum mechanics is the big black box that resides right at the heart of fundamental physics, a problem that eventually is more troubling than even dark energy, the Higgs mass, or other fine-tuning problems. By looking into the relations of entanglement and gravity, we arrive at the forefront of quantum gravity research. It is what Arkani-Hamed is calling the "central drama" of the twenty-first century.[38] And it has truly dramatic consequences for our understanding of space and time.

7

ONE BEYOND SPACE AND TIME

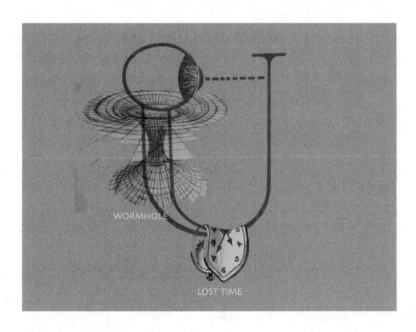

THE MONISTIC UNDERPINNING OF NATURE ENTAILED by quantum entanglement reveals to us a vast and entirely new territory to explore. It defines a new foundation of science and turns our quest for a theory of everything upside down—to build on quantum cosmology rather than on particle physics or string theory. But how realistic is it for physicists to pursue such an approach? Surprisingly, it is not just realistic—they are actually doing it already.

Researchers at the forefront of quantum gravity have started to rethink space-time as a consequence of entanglement. An increasing number of scientists have come to ground their research in the nonseparability of the universe. Hopes are high that by following this approach they may finally come to grasp what space and time, deep down at the foundation, really are.

Of Ants and Gods

The problem with gravity is that gravity is different. Albert Einstein's most important legacy is the revelation that gravity isn't a force like any other but is intimately connected with the inner workings of space and time. What eventually amounted to a complete liquidation of the classical concept of space and time as a fixed background on which physics would happen was indicated already in his theory of special relativity. With an obvious reference to Plato's cave allegory, Hermann Minkowski, the great mathematician and teacher of Einstein, summarized his student's theory thusly: "Henceforth space by itself, and time by itself, are doomed to fade away into mere shadows, and only a kind of union of the two will preserve an independent reality."[1] Indeed, Einstein had found that two observers moving at different speeds won't agree about the distance or time span between two events or even, in some cases, about which event happened first. Individually, time and space can be shrunk or stretched. Only space-time distances, the difference between time spans and distances squared, remain independent of the observer.

This defining feature of relativity has motivated physicists and philosophers to picture the universe as a four-dimensional "block" universe. In this context, time can be understood as a fourth dimension, with the important difference that we can go back and forth in space, but we must always advance in time. According to this view, instead of moving through space while time passes, subjects and objects move simultaneously through space and time. "There is no doubt that relativity . . . put in place a spatial view of time: time and space appear to be aspects of a single four-dimensional reality," confirms Oxford philosopher Simon Saunders.[2] As a consequence, space and time can then be depicted together—like the sequence of panels in a comic enabling

One Beyond Space and Time

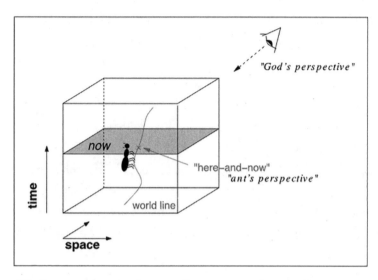

While an ant experiences time as passing as it moves along its lifeline, from an outside "God's perspective" onto space-time, the entire life of the ant is exposed.

the reader to view various events in the protagonist's biography at once. What is experienced as temporal from within (for the protagonist) can be understood as an unchanging path through space and time from a hypothetical outside perspective (such as for the reader).

The debate over whether nature is temporal or timeless is not a new one; it has its origins in ancient times. Around 520 BCE, Parmenides and Heraclitus developed competing worldviews on this very subject. Whereas Heraclitus argued that "no man ever steps in the same river twice," Parmenides asserted that "change is an illusion." Despite these seemingly oppositional worldviews, both philosophers agreed that the universe is an all-encompassing unity—that on the most fundamental level, "all is One."

The worldviews of Heraclitus and Parmenides thus can be seen to correspond to complementary perspectives rather than contradictory ontologies. Physics Nobel Prize winner Frank Wilczek has described these different perspectives as the "God's eye" (the outside view onto the entirety of space-time) versus the "ant's eye" (the individual experience along a certain path through space-time). "A recurring theme in natural philosophy is the tension between the God's-eye view of reality comprehended as a whole and the ant's-eye view of human consciousness, which senses a succession of events in time," Wilczek explains. "Since the days of Isaac Newton, the ant's-eye view has dominated fundamental physics. We divide our description of the world into dynamical laws that, paradoxically, exist outside of time, and initial conditions on which those laws act." Wilczek believes that it is time for a change in that point of view: "That division has been enormously useful and successful pragmatically, but it leaves us far short of a full scientific account of the world as we know it." Predicting the future of physics in 2016, Wilczek wrote, "To me, ascending from the ant's-eye view to the God's-eye view of physical reality is the most profound challenge for fundamental physics in the next 100 years."[3] In a talk about the same topic given at Brown University a few months earlier, Wilczek associated the God perspective with the timeless philosophy of Parmenides and Plato and the ant's view with Heraclitus.[4] And he quoted Erwin Schrödinger's friend Hermann Weyl's portrayal of the implications of this conception: "The objective world simply is, it does not happen. Only to the gaze of my consciousness, crawling along the lifeline of my body, does a section of this world come to life as a fleeting image in space which continuously changes in time."[5]

Wilczek's ant and God perspectives recall the frog and bird perspectives of his Massachusetts Institute of Technology (MIT) colleague Max Tegmark, which contrasted the quasi-classical experience of a local quantum observer with the hypothetical observer looking at the entire quantum universe from outside. From the bird perspective, there is no environment triggering decoherence; thus, presumably, the universe will be experienced as a single quantum object. It is tempting to associate the ant with the frog perspective and the bird with the God perspective. But how are space and time related to quantum mechanics? This puzzle isn't completely solved—yet—but to most physicists, it is clearly related to the extremely challenging task of developing a quantum version of general relativity: Einstein's theory of space, time, and gravity.

Sadly, Minkowski didn't live on to see this next coup of Einstein's, a theory that completely abandoned space and time as a rigid stage and made it a player in its own right. Gravity, Einstein had realized, wasn't a force in the traditional sense: for an observer in free fall, there would be no effect of gravity—he would feel weightless (until he hit the ground, of course)—while an observer in an upward-accelerating container would experience a pull toward the ground that would feel just like gravity. Gravity, Einstein discovered, was nothing but a pseudo force, just like the centrifugal force that seems to push an observer outside when he is entering a turn but is basically nothing but the consequence of the observer's inertia, which tends to keep him on a straight trajectory. But then, if gravity isn't a force in its own right, what makes the moon orbit Earth and Earth orbit the sun? According to Einstein, mass and energy warp space and time, and the resulting curvature of space-time forces the moon to follow what would correspond to a straight trajectory in flat space but looks like a circle in the warped geometry around Earth. As a consequence, space and time can now curve, warp, and oscillate, like the membrane of a drum: "Space-time tells matter how to move; matter tells space-time how to curve," John Wheeler famously summarized Einstein's abstract and complicated theory. As Wheeler's bon mot points out, space-time and matter aren't independent of each other: the first part of the witticism describes Einstein's equations, the backbone of general relativity describing how mass and energy bend space-time and determine its geometry.

But there remains a problem with Einstein's view that gravity is reducible to the warping of space and time. No proper quantum mechanical description has been found for this relationship. We know that this can't be the full story, since what is warping space and time, according to Einstein, are the masses and energies of stars and planets. These objects ultimately consist themselves of particles and fields, or—more properly—quantum fields and their particle-like excitations.

So Einstein's equations, relating quantum mechanical mass and energy on one side to the classical geometry of space and time on the other side, are a useful approximation but can't be fully correct, unless spacetime geometry is quantum as well. A quantum theory of gravity, space, and time is needed, and this problem occupied Wheeler probably more than anything else: "Even as I was learning general relativity for the first time by teaching it in the spring of 1953," he remembered in his autobiography, "I was pondering its link to quantum theory."[6]

A Wave Function for the Universe

In 1965, when John Wheeler had to change planes at the Raleigh-Durham airport, he called Bryce DeWitt, who worked at the nearby University of North Carolina and asked to meet him there.[7] DeWitt, who had co-organized the 1957 Conference on the Role of Gravitation in Physics, where Hugh Everett's many worlds interpretation was first discussed publicly, and who later became Everett's greatest champion, shared Wheeler's obsession with quantum gravity. While waiting with Wheeler for his connecting flight, DeWitt proposed to treat gravity just like Schrödinger had treated the hydrogen atom back in 1925. Wheeler, desperate to develop a theory of "quantum gravidynamics" (or, more common today, "geometrodynamics"), in analogy to his famous student Richard Feynman's quantum electrodynamics, passionately encouraged DeWitt to do so. And DeWitt did when, not much later, he spent some time at the Institute for Advanced Study neighboring Wheeler's university in Princeton, where he had further occasions to discuss his ideas and progress with Wheeler. The result was a Schrödinger-type equation for the quantum mechanical wave function of the universe—with one notable difference: where Schrödinger's equation determined how the wave function of an electron would develop in time, in DeWitt's version, known as the Wheeler-DeWitt equation, there was a zero. The wave function of the universe, it seemed, described a nonchanging or static, timeless universe: "Nothing ever happens in quantum gravidynamics," DeWitt concluded.[8] "Quantum theory can never yield anything but a static picture of the world."[9]

The reason for the disappearance of time was similar to the disappearance of electron orbits in usual quantum mechanics that had already bothered Werner Heisenberg in his epic struggle on Helgoland. As a consequence of the wavy nature of the electron, its position and momentum couldn't be determined simultaneously with arbitrary accuracy. Now that gravity (i.e., the curvature of space-time) had to be quantized, space and time itself had to be bestowed with a wavy nature. In the quantum version of general relativity, as Wheeler had conjectured ten years earlier, space-time may dissolve into a foamy structure "made up not merely of particles popping into and out of existence without limit, but of space-time itself churned into a lather of distorted geometry."[10] This would entail dramatic consequences. "So great would be the fluctuations that there would literally be no left and right, no before and no after. Ordinary ideas of length would disappear. Ordinary ideas of time would evaporate," Wheeler expected.[11] To obtain something playing the role of location and momentum in this context, DeWitt had decomposed four-dimensional space-time into a stack of three-dimensional spaces, just like a stack of cards. Location in standard quantum mechanics became the internal curvature of space (as if the picture on a card were warped), and momentum became the external curvature in the fourth dimension (as if the card were bent). Time was like a counter or scale for the succession of possible spaces in the stack.[12] Now remember that in standard quantum mechanics, the electron trajectory that describes the ordered succession of locations or events of a particle disappears, while individual locations or events remain. In the same way DeWitt's stack of spaces dissolves into individual spaces in his approach to quantum gravity, and time disappears, as it signifies (or "parametrizes") the ordered succession of spaces. As a consequence, "with spacetime churned into quantum foam, space and time in fact lose their meaning. When we blend the two greatest theories of the twentieth century, quantum theory and general relativity, we have to conclude that time is a secondary concept, a derived concept," Wheeler explains.[13]

When Niels Bohr had developed his interpretation of Schrödinger's wave function, he adopted a preexisting, classical measurement apparatus. As DeWitt soon realized in his effort to quantize gravity, this strategy couldn't be employed: "Here . . . the whole universe is the object of inspection; there is no classical vantage point, and hence the interpretation question must be reargued from the beginning."[14] This brought DeWitt back to Everett's interpretation: "Everett's view is a very natural one . . . It is possible that Everett's view is not only natural but essential."[15] But the obvious problem remained: How could one make sense out of a universe without time? And for that matter, could time be regained?

In Search of Lost Time

Julian Barbour's revelation about the unreality of time came along with a headache. Waking up under a star-spangled sky on an early October morning in the Bavarian Alps, he immediately knew he wouldn't be able to make it to the summit: "I still remember vividly the brilliant stars of Orion and the other winter constellations, high in the sky before the October dawn. But stars or no stars, I could not face the climb with that headache."[16] Instead, while his friend Jürgen set off to climb Mount Watzmann, Barbour took two aspirin and went back to his bunk.

When he woke up an hour or two later, his thoughts went back to an article by Paul Dirac—who had developed the relativistic version of quantum mechanics—that he had read the day before during the train ride. In this article Dirac questioned the fundamental validity of Einstein's concept of space-time. "This prompted an even more fundamental question: what is time? Before Jürgen had returned, I was—and still am—the prisoner of this question," Barbour remembers.[17]

What Barbour had started to question was whether Feynman was right when he said, "Time is what happens when nothing else happens."[18] Instead, as Barbour mused, time may be "nothing but change," implying "that time does not exist at all, and that motion itself is pure illusion."[19] Wasn't time redundant already in classical physics? Wasn't it enough to describe motion in relation to other motion, such as the coincidence of a certain location of an object with a specific position of the pointer on a clock? "We do abstract time from motion," Barbour realized.[20] For thirty-five years Barbour pondered this notion of the unreality of time, until he condensed his thoughts into a popular book. Fittingly, he calls his timeless universe "Platonia" and refers to the Eleusinian mysteries along the way. *The End of Time*, judges Oxford philosopher Simon Saunders, "is gold" and "a masterpiece."[21] And yet he is still puzzled: "So time does not exist . . . But if there is no time, how do we account for the fact that it seems there is? . . . I do not know if it really makes sense."[22] The Wheeler-DeWitt equation provides the key to resolving this difficulty.

Initially, Barbour developed his thoughts without giving much thought to quantum mechanics. As of 1971, Barbour writes, he "had no thought of applications to quantum mechanics."[23] When he finally did, he found out that his "German physicist friend" H. Dieter Zeh and Zeh's student Claus Kiefer had arrived at very similar conclusions, albeit from a totally different direction.[24] Zeh and Kiefer had started with the timeless Wheeler-DeWitt equation and then tried to identify a parameter that could be understood as an "internal time," similar to the pointer states of a clock. "Firstly, and most important, is the necessity to derive the standard notion of time . . . as an approximate notion from the Wheeler-DeWitt equation," Kiefer realized.[25] But how exactly is time regained?

In Chapter 3 we got to know decoherence as an agent of classicality and a generator of matter. In the same way decoherence becomes now a generator of time. Indeed, for this task Zeh could benefit again from his amazement, which had already helped him to discover decoherence many years earlier, when he felt vexed about how a time-independent nucleus could be approximated as composed of time-dependent components. When Kiefer and Zeh applied a similar approach to DeWitt's timeless universe, they discovered that time would actually emerge once an observer focused only on a part of the fundamentally timeless whole. The process of decoherence would create an approximate, emergent time when irrelevant tiny density fluctuations or gravitational waves were neglected. The rest of the universe, observer included, would feature an emergent time parameter, they found. What's more, this "time illusion," as one may say, would inevitably point in the direction of cosmic expansion and would even come to a halt should, for some reason, the cosmic expansion stop. In this sense "the universe defines its own time," albeit as an emergent, nonfundamental phenomenon, experienced from a specific perspective.[26] "Among the consequences are the fundamental timelessness of quantum gravity, the approximate nature of a semiclassical time, and the correlation of entropy with the size of the Universe," explains Kiefer.[27] "We are able to understand from the fundamental picture of a timeless world both the emergence and the limit of our usual concept of time," he stresses, adding that "the Wheeler-DeWitt equation . . . may or may not hold at the fundamental Planck scale," the enormous energy where quantum gravity effects were assumed to become large.[28] "But as long as quantum theory is universally valid, it will hold at least as an approximate equation . . . In this sense, it is the most reliable equation of quantum gravity, even if it is not the most fundamental one."[29]

We end up with a bizarre quantum world. "In quantum gravity, the world is fundamentally timeless and does not contain classical parts," Kiefer summarizes.[30] Barbour agrees: "The quantum universe is static. Nothing happens; there is being but no becoming. The flow of time and motion are illusions."[31] "Forget Time," encourages also Carlo Rovelli, the author of the best-selling book *Seven Brief Lessons of Physics* and one of the fathers of "loop quantum gravity," string theory's major contender for a quantum theory of gravity.[32] While undeniably time for us is a basic experience, it isn't understood as a fundamental property of the universe anymore. Instead, time is in the eye of the beholder, a feature of our point of view onto the universe. While Rovelli admits that "the notion of time is extremely natural to us," he emphasizes that this holds true "only in the same manner in which other intuitive ideas are rooted in our intuition because they are features of the small garden in which we are accustomed to living," just as is the case for "absolute simultaneity, absolute velocity, or the idea of a flat Earth and an absolute up and down."[33]

The notion of a timeless universe is hard to swallow, even for the thinkers responsible for drawing it all up. "That sounds insane," wrote the philosopher Saunders, when he first encountered the timeless conception of the universe arrived at by Barbour, Kiefer, Zeh, and Rovelli.[34] And really, as Zeh points out, this straightforward implication of quantum gravity was mainly ignored: "Almost all scientists that contributed to the development of quantum gravity seem to have understood this aspect of timelessness (just as many other aspects of quantum theory) as entirely formal."[35] But is it really so surprising that quantum cosmology, striving for a fundamental description of the universe, is timeless? To answer that question, we return now to the last chapter's discussion about what "fundamental" really means. We return to entropy.

Entropy and the Seeds of Time

When we first introduced entropy, we pointed out that it is something like a measure of ignorance; or, expressed the other way around, the less entropy occurs in a description of nature, the less information about its inner workings is discarded and the more fundamental the description. The lack of entropy can be understood as a quantifier of fundamentality.

What we didn't mention at that point was entropy's intimate relation with time. According to the famous second law of thermodynamics, entropy is supposed to increase with time—unless it is already maximal or forced to decrease by increasing the entropy somewhere else even more. "The Second law of thermodynamics is usually regarded as the major physical manifestation of the arrow of time, from which many other consequences can be derived," Zeh, for example, emphasizes in an essay titled "Open Questions Regarding the Arrow of Time."[36]

Thus the direction of time can usually be identified from the increase of entropy: when we see a cup falling from a table and disintegrating on the floor, we aren't surprised (though we may be irritated). In contrast, when we see a pile of shards on the floor join to compose a cup and then jump up onto a table, we usually assume we are watching a movie in reverse, although the process is in principle possible. It is just ridiculously unlikely for the momenta of the air molecules around the broken cup to conspire in a way to hit the shards so that the cup first gets reassembled and then is pushed back up onto the table. The reason is again that there are many more ways (or "microstates") for things to be broken than for things to be undamaged: if you push a car down a cliff, you can expect to end up with a mess of screws, dented metal, and broken glass. But you will need many tries in dropping screws, dented metal, and broken glass to end up with an undamaged car. What we usually experience as the passage of time is the evolution of less probable toward more probable macrostates, which typically amounts to destruction and equilibration.

Often overlooked in this context is that entropy is subject to some arbitrariness, as it depends on the definition of your macrostate. If you look at the various conditions a car can have, depicted in the figure in Chapter 6, the consensus would ordinarily be to identify the situation in the upper-left image as "unbroken." This is, however, a consequence of the purpose most people associate with cars: the wish to drive it. If one wants to play soccer with the car's wheels or to cut something with a shard, a different microstate may as well (and more appropriately) be defined as "unbroken."

Now, since entropy increase characterizes the passage of time, this suggests that time itself is nothing else but entropy increase. In such a view, entropy constitutes what we perceive as the arrow of time. If, as argued above, entropy is a feature of a coarse-grained and somewhat arbitrary perspective onto the universe, and if the quantum universe as the fundamental description of nature has vanishing entropy, it thus appears likely that the quantum universe as a whole is timeless. This is exactly what has been found in quantum cosmology.

Observing the universe from the frog perspective leads to a non-vanishing entropy (in this perspective) and thus an emergent arrow of time. As Kiefer and Zeh write in a coauthored paper, "The emergence . . . of quasiclassical properties (including spacetime) from a wave function . . . relies on the most fundamental of all 'irreversible' processes . . . namely on decoherence." "Decoherence," they explain, "determines which kind of properties emerge in the form of a . . . branching of the wave function into specific 'world components' such as those with definite spacetime geometries."[37] In a quantum universe, time isn't fundamental but rather a consequence of our coarse-grained perspective onto our environment. Indeed, this connection had already been suggested by Everett. In a handwritten draft found by his biographer Peter Byrne in his son Mark Everett's basement, Everett stressed that "the apparent irreversibility of natural processes is understood also as a subjective phenomenon, relative to observers who lose information in an essential manner, still within a determinate framework which is overall reversible."[38] In the original version of his thesis, he clarified that such "irreversible phenomena . . . arise from our incomplete information concerning the system, not from any intrinsic behavior of the system."[39]

The emergence of time so far constitutes the only concrete derivation of a seemingly fundamental property of our classical universe by means of decoherence actually realized. It may thus serve as a proper example inspiring similar approaches to the emergence of other fundamental properties such as symmetries and conservation laws from a fundamental quantum description. But what about the intimate connection between space and time in Einstein's theory? If monism is right and there is only one thing in the universe, what sense does it make to speak about its location? A location in relation to what? Indeed, cutting-edge research on quantum gravity now seems to suggest that space, just as time and matter, may be not fundamental but emergent as well—based on insights that grew out of the struggle to understand some of the strangest beasts in the universe: black holes.

The Mysteries of Black Hole Information Processing

Black holes rank among the most outlandish predictions of general relativity. This is not to say that they are rare: each star with a mass larger than three solar masses ends its life as a black hole, and virtually all galaxies in the universe, including our own Milky Way, feature gigantic black holes of many millions and even billions of solar masses at their core.

But what are black holes? And why does Einstein's theory imply their existence? When Einstein published his equations of general relativity in 1915, he first believed them to be unsolvable. But less than two months later, the German Jewish physicist Karl Schwarzschild came up with a solution—an achievement that appears even more remarkable as Schwarzschild at this time was fighting as a volunteer in World War I at the German-Russian front. As a German Jew and civil servant, Schwarzschild was eager to prove himself a true patriot. Sadly, only a few months later, Schwarzschild came down with a rare autoimmune disease, was hospitalized as a disabled veteran, and died within weeks. Even worse, maybe, in hindsight, Schwarzschild's sacrifice wasn't honored. Twenty years later his children were cast out of the country, and one of his sons committed suicide during the persecution of Jews in Nazi Germany.

Schwarzschild's solution described the gravitational field around a spherical mass, such as a star or planet. But the solution had an interesting feature: when the spherical mass was smaller than a certain value, the so-called Schwarzschild radius, it formed a black hole. What makes black holes special is the extreme warping of space and time they feature: on the Schwarzschild radius, gravity gets so strong that everything, even light, is pulled inside the black hole. For what falls into a black hole, there's no way back; starting from the Schwarzschild radius, the black hole becomes a one-way street. From that observation, a black hole seems to be a rather complex object, made up of all the clutter that has been sucked inside. Surprisingly though, the contrary is the case: Schwarzschild's black hole is characterized by its mass only. Later, variants of black holes had been discovered that featured electric charge and rotation, but that's it. Black holes are entirely specified by their mass, charge, and angular momentum. "A black hole has no hair," John Wheeler and his students Charles Misner and Kip Thorne wrote metaphorically of this finding in their famous textbook on gravitation.[40]

As a consequence, what falls into black holes is lost—at least that's what people thought in the 1970s. And at least in principle, this lost information can be understood as entropy. In fact, in 1973, Wheeler's PhD student Jacob Bekenstein noticed that "there are a number of similarities between black hole physics and thermodynamics."[41] Entropy, for example, could be linked to the area of the black hole horizon, Bekenstein mused. This hypothesis was based on a result by Stephen Hawking three years earlier: just like entropy, following the second law of thermodynamics, "the area of the horizon cannot decrease."[42] "When the black hole interacts with anything else the area will always increase," Hawking explained.[43] Still, Hawking was reluctant to take the analogy between entropy and a black hole's surface area literally. After all, a black hole having entropy should also feature a temperature and thus radiate, which was deemed absurd. "Bekenstein," Hawking felt, "had misused my discovery."[44] Hawking changed his mind only after he had analyzed quantum fields in the vicinity of the black hole horizon and discovered that, contrary to expectation, black holes indeed emit radiation. Hawking's result is a special case of the more general Unruh effect mentioned in Chapter 3. This phenomenon, named after a 1976 paper by Wheeler's student William Unruh, implied that an accelerated observer would see particles in the vacuum. As according to general relativity, gravity can be understood as acceleration, any observer trying to escape a black hole is expected to see a stream of particles pouring out of it. These particles originate from the vacuum in quantum field theory, which isn't conceived of as empty but rather as a hot soup of "virtual" particles popping in and out of existence. In the extreme gravitational field close to the black hole horizon, such quantum fluctuations can suck energy out of the black hole and thereby transform their virtual particles into real radiation.

But Hawking's discovery only made black holes more mysterious. After eating stars and other astronomical objects like a horse, the galactic monsters were supposed now to dissolve into radiation and eventually disappear. This triggers the question about what would happen to all the information that had fallen into the black hole, the specific properties and characteristics of all the stars and other things that disappeared behind the black hole's horizon. Hawking believed it would simply disappear—a solution that is at odds with quantum mechanics, however: just as nothing happens in physics without a cause, information cannot simply disappear into thin air.

It took until the mid-1990s for a solution to the dilemma to be proposed by Gerard 't Hooft and Leonard Susskind. The proposal was based on a feature of black holes that often had been ignored. While a black hole is usually described as insatiably swallowing everything that falls in its direction, and while this is indeed what an observer falling into the black hole experiences, an observer remaining at rest outside the black hole witnesses an entirely different story. From her perspective, the strong gravitational field on the horizon bends spacetime so strongly that time comes to a halt. From the perspective of this observer, thus, nothing ever falls into the black hole, and things seen falling toward the black hole freeze on its horizon instead.

Who, then, is correct? The answer 't Hooft and Susskind came up with was Solomonic, inspired by Bohr's notion of complementarity: maybe both of them. If, according to Bohr, an electron can be both a particle and a wave depending on how it is observed, maybe things can be both inside the black hole and on its horizon. For the outside observer, information never has entered a black hole and can leave it again with the Hawking radiation emitted during the black hole's evaporation process. The inside of the black hole would then be something akin to a hologram: a three-dimensional image that can be understood as being produced from information stored on the two-dimensional horizon.

For eighteen years, this story about "black hole complementarity" and the "holographic principle" remained the conventional wisdom. It was reinforced when, in 1997, the idea got unexpected support from string theory. At that time, researchers tried to make sense out of the relation among various versions of string theories that had been discovered and threatened the status of string theory as the unique candidate for a theory of quantum gravity. They did so by searching for relationships that would translate or map the various theories onto each other. One such relationship proposed by Argentinian Juan Maldacena suggested a correspondence between two toy universes, one with gravity and one without, featuring different numbers of spatial dimensions. What blew his peers' minds was that Maldacena's conjecture seemed to imply that a complete, spatial universe with gravity could be equally well viewed as a hologram spawned by a quantum field theory defined on its volume's boundary surface. (More concretely, Maldacena assumed a negative vacuum energy, implying a geometry known as Anti-de Sitter (AdS), and related it to a specific quantum field theory known as conformal field theory (CFT). As a side effect, the relationship allowed physicists to jump back and forth between the two descriptions, and problems deemed intractable in one description turned out to be feasible in the other. Just like black holes, Maldacena's discovery suggested, universes could be understood as holograms. In turn, black holes could be used to model universes (in a much less complex way, that is). Maldacena's paper was soon adopted as the prime example of the holographic principle and quickly became the most popular paper in the history of high-energy and particle physics, accumulating more than twenty thousand citations up to 2020. What was known as the "black hole information paradox" seemed to have been solved. Black holes could be defined by their horizon, and what happened inside black holes, it seemed, could be ignored.

Then, in 2012, another problem with quantum mechanics emerged. As Ahmed Almheiri, Donald Marolf, Joseph Polchinski, and James Sully realized, in order to transport information with Hawking radiation, its quanta would have to be entangled among each other, but they also had to be entangled with that part of the original quantum fluctuations that was sucked into the black hole. But entanglement has to be "monogamous," quantum physicists found: particles can be perfectly correlated only with one partner, not with two or more. To clarify the problem, Polchinski compared black hole evaporation with burning coal, which produced heat irradiated by photons: "The early photons are entangled with the remaining coal," but ultimately all information about the burning piece comes out with the radiation.[45] As Polchinski points out, "The burning scrambles any initial information, making it hard to decode, but it is reversible in principle."[46] Now "a common initial reaction to Hawking's claim is that a black hole should be like any other thermal system," as Polchinski explains, "but there is a difference: the coal has no horizon."[47] As Polchinski emphasized, "The early photons from the coal are entangled with excitations inside, but the latter can imprint their quantum state onto later outgoing photons. With the black hole, the internal excitations are behind the horizon, and cannot influence the state of later photons."[48] Thus, to get the infalling information out again and keep information preserved, the entanglement between infalling and outgoing quantum fluctuations must be cut somehow. The consequence was "drama," as the authors described it: a huge release of energy interpreted as a "firewall" at the horizon. Such a firewall on the horizon would totally invalidate Einstein's original starting point though: the observation that gravity was imperceptible during free fall. The physics of black hole information processing seemed to be back at square one.

New ideas were needed, and they would end up coming from a reconsideration of the relationships between space-time and quantum mechanics.

ER=EPR?

The firewall paradox results from a clash between particle locations in black hole space-times and quantum entanglement. Thus researchers have recently started to speculate as to whether space-time and entanglement, usually understood as independent concepts, may feature a hidden relation. If this is the case, the nonlocality inherent in quantum entanglement may be inherited by space-time, with dramatic consequences for our understanding of space itself.

One such connection between black holes and quantum nonlocality is an idea that is known by the buzzword "ER=EPR." "EPR" here refers to the famous Einstein-Podolsky-Rosen publication on quantum entanglement. "ER" refers to a paper by Einstein and Nathan Rosen from the same year, 1935, discussing wormhole-like solutions in general relativity.

Wormholes arise in generalized black hole space-times. While observers at rest outside the black hole never see anything fall into it, a free-falling observer can cross the horizon without any problem but can never come back. This is surprising, since the Schwarzschild space-time describing black holes doesn't single out a specific direction of time. Thus, for each particular solution singling out a time direction (such as the ones allowing an observer to fall into a black hole but not to escape from it), there should exist a corresponding time-inverted solution. In other words, if there exists a one-way street into the black hole, crossing the Schwarzschild radius from outside, there should also be another one-way street crossing the Schwarzschild radius from the inside. Indeed, while nobody can escape a black hole after crossing the horizon, the space-time describing black holes can be extended for arbitrary observers in a way that allows crossing the Schwarzschild horizon in an outward direction, only the horizon now connects the region outside the black hole to a hypothetical white hole (one that can be escaped but never be entered) or to a portal from a parallel universe. As Einstein and Rosen had shown briefly after they came up with the EPR paradox, this parallel universe is connected to our own universe by an "Einstein-Rosen bridge" or—in modern terminology—a "wormhole." Wormholes later became popular in science fiction as devices for faster-than-light interstellar travel (a kind of natural teleportation widget) or even time travel to the past; yet the original wormhole wasn't traversable: nobody, not even a particle, could actually pass through it.

Wormholes and entanglement indeed have one thing in common: "Locality appears to be challenged both by quantum mechanics and by general relativity," as Maldacena and Susskind argue.[49] Yet the identity "ER=EPR," first used in an e-mail from Juan Maldacena to Leonard Susskind looks—at first glance—like a conglomerate of nonsense. While entanglement describes the holistic properties of quantum systems ensuing from nonlocal correlations between their constituents, wormholes describe handles in space-time that connect faraway regions of the universe. The first is a common quantum phenomenon that doesn't allow for faster-than-light information exchange; the second is a property of a purely hypothetical, exotic space-time geometry that has been advocated as a tool for superluminal trips and time travel and usually is considered to be unstable or incompatible with the laws of physics. Susskind doesn't make it better by claiming that "ER=EPR" implies that "Everett=Copenhagen," ignoring that Everett's and Bohr's interpretations of quantum mechanics are based on absolutely contradictory notions of reality.

Yet, the ER=EPR conjecture gains in plausibility once one approaches it from the parallels between Everett's many worlds and the multiverse that is predicted in cosmic inflation and married with the string theory landscape. As mentioned in Chapter 6, researchers, including Anthony Aguirre, Max Tegmark, Yasunori Nomura, Raphael Bousso, and Leonard Susskind, had advocated that Everett's parallel realities and the "bubble" or "baby" universes expected to pop up in eternal cosmic inflation and surmised to accommodate all possible varieties of the laws of physics allowed by string theory are one and the same thing—in the sense that both exhaust the space of possible realities. Now, while any link between different baby universes could be realized only with the help of wormholes (i.e., shortcuts through space-time), quantum superpositions such as particles being "half-here" and "half-there" or Schrödinger's zombie cat do indeed connect different Everett branches or "worlds." And superposition, as we have seen, can be understood as a specific form of entanglement, as entanglement in occupation space. So if wormholes connect different baby universes, and entanglement links different Everett branches, and both are the same, maybe wormholes and entanglement could indeed be related.

Maldacena and Susskind boldly claimed that "any pair of entangled black holes will be connected by some kind of Einstein-Rosen bridge."[50] What's more, for Maldacena and Susskind, the wormhole "is a manifestation of entanglement," meaning that black holes connected by wormholes are entangled, and entangled black holes are connected by wormholes.[51] They "go even further and claim that even for an entangled pair of particles, in a quantum theory of gravity there must be a Planckian bridge between them, albeit a very quantum-mechanical bridge which probably cannot be described by classical geometry."[52] In quite general terms, then, any entangled quantum system can be understood as connected by a wormhole and vice versa. These wormholes may help to avoid the notorious firewalls on the black hole horizon, as they connect the black hole interior to the Hawking radiation outside. The particles inside and outside the black hole could be considered one and the same, just like a star could be seen twice, once directly and once through the wormhole. "By the ER=EPR principle," Maldacena and Susskind explain, the black hole "is connected to the outgoing radiation by a complex Einstein-Rosen bridge with many outlets"—a picture that has been compared to an octopus: "The outlets connect the bridge to the black hole, and to the radiation quanta."[53]

Susskind doesn't stop there. He generalizes this finding to a general relationship between general relativity and quantum mechanics. "What all of this suggests to me, and what I want to suggest to you," Susskind writes elsewhere, "is that quantum mechanics and gravity are far more tightly related than we (or at least I) had ever imagined. The essential nonlocalities of quantum mechanics parallels [sic] the nonlocal potentialities of general relativity."[54] According to Maldacena and Susskind, entanglement and wormholes are the same. "If there is a lesson to be gained from recent work on quantum gravity it is that geometry and quantum mechanics are so inseparably joined that each may not make sense without the other," Susskind concludes.[55]

Physicists are divided over the question of how literally these wormholes are to be taken. But there seems to be consensus that they indicate a nonlocal property of space-time. As of 2020, physicists, employing wormholes and baby universes emerging inside black holes, are again optimistic that the black hole information paradox is close to being solved and that no information is fundamentally lost.[56] Finally, the ER=EPR conjecture resonates well with a set of papers by different authors that appeared in 2009 and 2010. While these works seem to have originated entirely independently, they have at least two things in common: first, all of them were considered so far-fetched that the authors had difficulties getting them accepted for publication; second, all of them proposed that space-time isn't fundamental but sewn together by entanglement.

Space-Time from Entanglement

In January 2010, Erik Verlinde at the University of Amsterdam in the Netherlands compared the gravitational pull of a black hole to osmosis: the diffusion of a liquid through a membrane driven by entropy. His approach was based on work from 1995 by American Ted Jacobson at the University of Maryland. Jacobson had turned the logic of black hole thermodynamics upside down. Starting from the relations developed by Hawking, Bekenstein, and others in the 1970s, linking the astrophysical monsters to the physics of heat and steam, Jacobson was able to derive the Einstein equations, the backbone of general relativity, from thermodynamics. Fifteen years later Verlinde went a step further and argued that gravity could be understood as an "entropic force," a force entirely created by entropy.

Such entropic forces are macroscopic effects driven by the tendency to maximize entropy that don't have any microscopic equivalent. Osmosis, for example, pulls liquids against the force of gravity through membranes in order to balance the concentrations of substances dissolved in these liquids. Now gravity, Verlinde proposed, may be an entropic force itself. Gravity can, according to Verlinde, "emerge from a microscopic description that doesn't know about its existence."[57] To underpin his proposal, Verlinde starts with two observations: first, that "of all forces of Nature gravity is clearly the most universal," as it "influences and is influenced by everything that carries an energy, and is intimately connected with the structure of space-time," and second, that the laws of gravity "closely resemble the laws of thermodynamics."[58] Thermodynamics had been understood in terms of information and entropy in the nineteenth century. Thus Verlinde sets out to derive gravity from information: "Changes in . . . entropy when matter is displaced lead to an entropic force, which . . . takes the form of gravity. Its origin therefore lies in the tendency of the microscopic theory to maximize its entropy."[59] A crucial assumption that Verlinde adopts from the holographic principle in black hole physics and string theory is that there is only a finite amount of information corresponding to a given spatial volume. "Space is in the first place a device introduced to describe the positions and movements of particles. Space is therefore literally just a storage space for information," Verlinde explains.[60]

With these ingredients, the black hole horizon can become what the membrane is for osmosis, and the laws of gravity can be derived from the increase of entropy when a bit of information merges with the black hole's horizon. While "the holographic principle has not been easy to extract from the laws of Newton and Einstein, and is deeply hidden within them," Verlinde summarizes, "conversely, starting from holography, we find that these well known laws come out directly and unavoidably."[61] The result is "the end of gravity as a fundamental force," and "if gravity is emergent, so is space time geometry," Verlinde concludes.[62]

Meanwhile, Mark Van Raamsdonk of the University of British Columbia in Vancouver, Canada, who had pondered similar ideas, was struggling to publish his own results. The trouble was that his paper kept getting turned down by the major journals, with referee reports implying nothing less than that he was a crackpot.[63] Eventually he succeeded when he submitted a shorter version of his original paper to the annual essay contest run by the Gravity Research Foundation and won. The first prize, awarded in May 2010, came along with publication in *General Relativity and Gravitation*, one of the journals that had rejected his paper previously.

Like Verlinde, Van Raamsdonk built his argument on the holographic principle, or Maldacena's conjecture, to be exact. But instead of studying the gravitational pull exerted by a black hole's entropy, he started with the entangled state of two particles. Naively, according to the AdS/CFT conjecture connecting a gravitating AdS space to a nongravitational quantum field theory, or CFT, such a state in the CFT corresponds to a superposition of two disconnected space-times—a superposition of space-times featuring, to be exact, the geometry discovered by Schwarzschild in 1915, since particles are spherically symmetric. Yet the expression Van Raamsdonk arrived at was a single space-time with two black holes connected by a wormhole. The entanglement between the particles creates, so it seems, a connection between their respective space-times. The reason for this unexpected result is that an observer in either one of the original two space-times in the superposition would see a single black hole embedded in a flat space-time from the outside, with its Hawking radiation originating from ignoring the other space-time—corresponding to decoherence in the quantum picture. Van Raamsdonk arrives at what he calls "a remarkable conclusion": "the state . . . which clearly represents a quantum superposition of disconnected spacetimes may also be identified with a classically connected spacetime."[64] Excitedly, he realizes that "we have glued together the corresponding geometries!"[65] In other words, "the emergence of classically connected spacetimes is intimately related to . . . quantum entanglement," or "classical connectivity arises by entangling."[66]

After that, Van Raamsdonk set out to perform the converse task: disconnecting a connected space-time by tuning down the entanglement in its quantum field theoretic equivalent. For that, he relied on a generalization of the relationship between black hole area and entropy that was discovered by Shinsei Ryu and Tadashi Takayanagi in 2006 when the two were postdocs at the University of California, Santa Barbara.

Instead of confining themselves to black hole horizons, Ryu and Takayanagi employed the AdS/CFT correspondence to explore more universally the relationships between areas and the entropy associated with entanglement. In particular, they had studied what happens to the interface of two spaces when the entanglement between the corresponding quantum field theories is decreasing slowly. They found that the area of the interface shrinks. What's more, Van Raamsdonk also looked at the mutual information the two spaces convey about each other. He discovered that for decreasing entanglement, this information gets suppressed as if it were transmitted by a heavy particle propagating over a long distance. Entanglement corresponds to proximity, Van Raamsdonk concluded, and zero entanglement between two quantum field theories effectively pinches off the connection between the corresponding space-time regions: "We can connect up spacetimes by entangling degrees of freedom and tear them apart by disentangling. It is fascinating that the intrinsically quantum phenomenon of entanglement appears to be crucial for the emergence of classical spacetime geometry," he concludes.[67]

A few weeks before Van Raamsdonk had posted his paper to the high-energy theory section on the internet platform arXiv, Brian Swingle, a PhD student at MIT, had arrived at essentially the same conclusions, posted to the arXiv's "Strongly Correlated Electrons" section.

Using methods typically employed in solid-state physics, Swingle concluded that "entanglement is the fabric of space-time," as he told science writer Jennifer Ouellette.[68] In other words, "you can think of spacetime as being built from entanglement."[69] And Van Raamsdonk agrees: "Spacetime . . . is just a geometrical picture of how stuff in the quantum system is entangled."[70]

Van Raamsdonk's and Swingle's work clearly struck a nerve. In 2015, the Simons Foundation launched the initiative "It from Qubit," described as "a large scale effort by some of the leading researchers in both communities [fundamental physics and quantum information theory] to foster communication, education and collaboration between them, thereby advancing both fields and ultimately solving some of the deepest problems in physics."[71] In this approach, space as we know it is understood as a consequence of entanglement, and quantum information science is employed to explain what string theorists and other researchers had found out about gravity. Around the same time, as pointed out in the "It from Qubit" proposal, the number of preprints in the arXiv's high-energy theory section with the term "entanglement" in the title exploded, following what looks like exponential growth. Entanglement—correlations originating from oneness—has replaced length "as the world-making relation," observes Norwegian physicist and philosopher Rasmus Jaksland.[72]

More Ideas for Physics Without Space

These sweeping developments that seem to uncover more and more links between entanglement and an emerging, nonfundamental spacetime are just a sampling of the ideas seeking to strip physics of its attachment to space and time. They contribute to mounting evidence for a monistic foundation beyond space and time.

One such idea proposed by Nima Arkani-Hamed and Jaroslav Trnka in 2013 derives from particle physics. The authors found a novel way to efficiently describe particle interactions with the help of an abstract geometrical object, the "amplituhedron," which has been described as resembling a "jewel in higher dimensions."[73] The method provides an alternative to the Feynman diagram approach usually employed in quantum field theories where particle scattering is depicted with tree-like diagrams of particles propagating through space-time, even if particles on internal lines are not observable but rather represent potential possibilities for starting from the initial and arriving at the final state. Instead, the authors introduce the amplituhedron, "a new mathematical object whose 'volume' directly computes the scattering amplitude," the basis for determining the probability of a particle reaction.[74] The approach abandons both notions of unitarity—the concept that nothing disappears into thin air—and locality, cornerstones of quantum field theories that no longer make sense in a universe where space and time dissolve. While the amplituhedron doesn't describe gravity yet, Arkani-Hamed is optimistic that it may help to elucidate how space and time emerge from a description beyond space and time. In his approach, Arkani-Hamed tells science writer Natalie Wolchover, "we can't rely on the usual familiar quantum mechanical space-time pictures of describing physics . . . We have to learn new ways of talking about it. This work is a baby step in that direction."[75]

Another idea that exploits potential quantum possibilities from a different viewpoint is advocated by Oxford physicists David Deutsch and Chiara Marletto: "the science of can and can't," as Marletto describes it.[76] "Constructor Theory," as characterized on the research group's website, "is a new approach to formulating fundamental laws in physics. Instead of describing the world in terms of trajectories, initial conditions and dynamical laws, in constructor theory laws are about which physical transformations are possible and which are impossible, and why."[77]

Constructor theory thus aims to rewrite the laws of physics in terms of statements about what's possible and what's not. The term "constructor" here refers to a universal production process that has been compared to an "all-purpose 3D printer, capable of constructing any physical object," a description Marletto agrees with.[78] She envisions how the framework can be extended to a theory of computation or a definition of life: "The universal constructor has all the physically allowed computations in its own repertoire, which means it is a universal computer, too," and "the physics of life would be considered a subpart of this more general theory of the universal constructor," Marletto explains.[79]

The idea is clearly inspired by quantum mechanics where Everett's universal wave function represents the collection of everything that is possible. For Deutsch, an early champion of Everett's interpretation, promoting the feature of possibility to the foundational principle in physics and beyond, coming up with constructor theory thus must have been an obvious step. As Deutsch had emphasized before, to understand "probability, and time, and the nature of existence and non-existence, and the self, causation, laws of nature, the relationship of mathematics to physical reality—for any of those, you've got to understand the multiverse." "Learn to think in terms of it, build on that understanding, and apply it to learning about everything else," he encouraged in 2010.[80]

Finally, a more directly monistic approach has been proposed by mathematician Michael Freedman, the 1986 recipient of the Fields Medal, often called the mathematicians' surrogate for the Nobel Prize.

"The purest question 'Why is there something instead of nothing?' seems out of reach; we, instead, attempt an answer to: 'Why does there appear to be a multitude of things?'" Freedman announced in his colloquium talk, titled "The Universe from a Single Particle," which he gave on July 15, 2021, at the prestigious Aspen Center for Physics.[81] In this scenario Freedman and his collaborator, Modj Shokrian Zini, try to break the symmetry of single-particle quantum mechanics in a way that arrives at a universe made up of interacting components. "That single particle . . . could . . . be the beginning of the story," Freedman and Zini write in a 2020 preprint.[82]

* * *

Whether space is stitched together by entanglement, physics is described by abstract objects beyond space and time or the space of possibilities represented by Everett's universal wave function, or everything in the universe is traced back to a single quantum object—all these ideas share a distinct monistic flavor. At present it is hard to judge which of these ideas will inform the future of physics and which will eventually disappear. What's interesting is that while originally ideas such as AdS/CFT were developed in the context of string theory, the ideas about holography and nonlocal quantum gravity seem to have outgrown string theory, and strings play no role anymore in the most recent research. A common thread now seems to be that space and time are not considered fundamental anymore. Contemporary physics doesn't start with space and time to continue with things placed in this preexisting background. Instead, space and time themselves are considered products of a more fundamental projector reality. Nathan Seiberg, a leading string theorist at the Institute for Advanced Study at Princeton, is not alone in his sentiment when he states, "I'm almost certain that space and time are illusions. These are primitive notions that will be replaced by something more sophisticated."[83] Moreover, in most scenarios proposing emergent space-times, entanglement plays the fundamental role. As Rasmus Jaksland points out, this eventually implies that there are no individual objects in the universe anymore; that everything is connected with everything else: "Adopting entanglement as the world making relation comes at the price of giving up separability. But those who are ready to take this step should perhaps look to entanglement for the fundamental relation with which to constitute this world (and perhaps all the other possible ones)."[84] Thus, when space and time disappear, a unified One emerges.

Conversely, from the perspective of quantum monism, such mind-boggling consequences of quantum gravity are not far off. Already in Einstein's theory of general relativity, space is no static stage anymore; rather it is sourced by matter's masses and energy. Much like the German philosopher Gottfried W. Leibniz's view, it describes the relative order of things. If now, according to quantum monism, there is only one thing left, there is nothing left to arrange or order and eventually no longer a need for the concept of space on this most fundamental level of description. It is "the One," a single quantum universe that gives rise to space, time, and matter.

"GR=QM," Leonard Susskind claimed boldly in an open letter to researchers in quantum information science: general relativity is nothing but quantum mechanics—a hundred-year-old theory that has been applied extremely successfully to all sorts of things but never really entirely understood.[85] As Sean Carroll has pointed out, "Maybe it was a mistake to quantize gravity, and space-time was lurking in quantum mechanics all along."[86] For the future, "rather than quantizing gravity, maybe we should try to gravitize quantum mechanics. Or, more accurately but less evocatively, 'find gravity inside quantum mechanics,'" Carroll suggests on his blog.[87] Indeed, it seems that if quantum mechanics had been taken seriously from the beginning, if it had been understood as a theory that isn't happening in space and time but within a more fundamental projector reality, many of the dead ends in the exploration of quantum gravity could have been avoided. If we had approved the monistic implications of quantum mechanics—the heritage of a three-thousand-year-old philosophy that was embraced in antiquity, persecuted in the Middle Ages, revived in the Renaissance, and tampered with in Romanticism—as early as Everett and Zeh had pointed them out rather than sticking to Bohr's pragmatic interpretation that reduced quantum mechanics to a tool, we would be further on the way to demystifying the foundations of reality.

Yet this is not the end of this road. While we have seen the potential implications for modern particle physics and quantum gravity, we must still ask, What about ourselves? How do we fit into a universe that is One, after all?

8

THE CONSCIOUS ONE

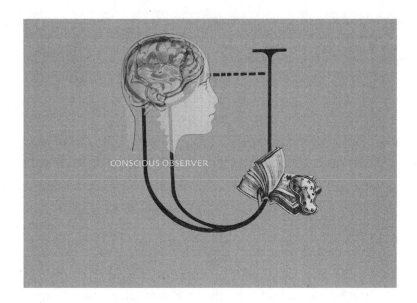

WE HAVE SEEN NOW HOW MATTER AND SPACE-TIME can emerge from a fundamental One, as a perspectival impression from an observer's point of view. Yet, up until this point we have kept silent about the observer herself. It is high time to change that. In order to find out how we ourselves, by simply looking at the universe, are creating it, we need to discuss how the universe is experienced consciously.

Setting Up the Screen

"I believe one can find many suitable images. I often have thought of Plato's allegory of the cave—we see only projections of reality," former

H. Dieter Zeh student and decoherence theory pioneer Erich Joos wrote to me by way of explaining how to picture the transition from quantum to classical.[1]

If Joos is right, we are the audience in nothing less than the greatest show in the universe, unfolding space, time, matter, and the entire cosmic history in every single moment, right in front of our eyes. Just as John Wheeler's U allegorizes, it is we who create the universe just by observing it—although "creating" here doesn't refer to a deliberate act. What we experience in our daily lives, or better, what we can experience at all, is an on-screen reality displaying our specific view on the underlying quantum reality. The story we live to see is a product of our characteristic perspective, right from our comfy chairs onto what's stored on the cosmic film roll. Enjoy the popcorn.

But how exactly does that happen? How do space, time, and matter emerge from an underlying "One"? We know that decoherence plays a crucial role in this process. When an observer—be it ourselves, our pets, a measurement apparatus, or a faraway alien—is observing something, he, she, or it gets entangled with the observed object. As we, as macroscopic beings, are in constant interaction with the rest of the universe, the information about quantum superpositions of the observed object leaks into our environment in next to no time. The result is a quasi-classical object with clearly defined specifications regarding the concrete property we are observing.

Sure enough, if we look at a quantum object's location, we indeed find it in a concrete place. But remember Niels Bohr's complementarity: a quantum object can be seen as a localized particle or as an extended wave (or anything in between). If, instead, we had looked for our object's velocity, we would find it as a plane wave stretched out over the universe, specified by a concrete momentum or wavelength.

So is it our choice what we see and what kind of world we live in? If yes, then why do all of us seem to agree about an objective reality? And why do all the objects around us seem to exist in defined places? Why do our addresses refer to locations instead of velocities or wavelengths? "Classical objects can be either here or there, but never both here and there. Yet, the principle of superposition says that localization should be a rare exception and not a rule for quantum systems," emphasizes Wojciech Zurek, who has contributed to the development of decoherence theory like probably no one else since the 1980s.[2]

This dilemma is known to quantum physicists as the "preferred basis problem": What decides whether we see particles or waves? Understanding decoherence is of some help here, but it doesn't solve the dilemma. In a nutshell, decoherence is a consequence of an interaction with the rest of the universe. If this interaction is local, if it happens at a specific place, then the quasi-classical reality decoherence produces will be local as well.

The reason we experience the things around us in concrete places is that we ourselves, our senses, and our measurement devices exist in concrete places. If my eye, my ear, or my lab equipment is situated in a concrete place, it can only interact locally with observed quantum objects and thus will produce a local on-screen reality of these objects. Indeed, whenever we measure velocities or wavelengths, we do that indirectly, by somehow translating these characteristics into spatial information. Eventually, the pointer on our measurement apparatus still has a well-defined position. Yet this is only half of an answer. We can go on and ask why this is the case: Why do we ourselves, our senses, and our lab devices exist in defined places rather than everywhere at once? And there are other open questions afflicting decoherence theory: What, for example, is the environment? Who decides how to split the universe into quantum system, observer, and environment?[3] Are other divisions possible? Could there, for example, be "quantum aliens" that aren't located in space but in wavelengths and frequencies and who are experiencing the universe as a conglomerate of spread-out waves instead of localized objects? This question, closely related to the preferred basis problem and known as the "quantum factorization problem," is "an open problem at the very heart of quantum mechanics," as Max Tegmark points out.[4]

There are theorists who have sought to strip decoherence of its subjective element. "Decoherence—the modern version of wave-function collapse—is subjective in that it depends on the choice of a set of unmonitored degrees of freedom, the 'environment,'" as Leonard Susskind and Berkeley physicist Raphael Bousso explain in a coauthored paper.[5] To solve this problem, the authors propose a concept called the "causal diamond," which provides a geometric description of which part of space-time is accessible to a local observer at all, given the fact that the speed of light defines a finite, limiting velocity for information flow. According to this idea, "environment" is at the very least what in principle can't be observed: "The causal diamond is the largest spacetime region that can be causally probed" and "leads to a natural, observer-independent choice of environment," Bousso and Susskind argue, to arrive eventually at "the global multiverse as a patchwork of decoherent causal diamonds."[6] Yet the individual causal diamond is still centered around a local observer.

The questions of what defines a local observer and why she is local in the first place are left unanswered. Foundational questions, such as "Where is our perspective onto the universe determined?" and "What defines the self of the observer?" remain to be resolved.

The Selfish Quantum

One potential answer holds that the universe itself sets up the screen and determines our perspective. This is Wojciech Zurek's position, and if he is right, decoherence is only part of the process of how the quantum becomes classical—which for Zurek is the same as "real." "Real" means, Zurek argues, that different observers will agree about a certain property of a quantum system. This is only possible if the information about the quantum state is recorded more than once. "We all monitor our world indirectly, eavesdropping on the environment. For instance, you are now intercepting a fraction of the photons scattered from this page. Anyone intercepting other fractions will see the same images," Zurek explains.[7] In his scheme, the environment becomes a "monitor" of the quantum system, and the information about the quantum system is stored redundantly in the memory of the environment.

To grasp what this means, we must remind ourselves once more how decoherence is described as working. First, we split the universe into quantum object, observer or apparatus, and environment. Then we adopt the frog perspective for the apparatus or observer, implying he, she, or it is ignorant about the exact state of the environment. This implies that—from his, her, or its perspective, quantum superpositions (such as a particle residing in different places at the same time) are suppressed, and the observed quantum object is perceived in a well-defined state having, for example, a specific location.

According to Zurek's scenario, the split of the universe into quantum object, observer, and environment is replaced by a split into quantum system and an environment consisting of many subsystems. "The environment is a witness to the state of the system," Zurek writes, and his scenario is defined by exactly this upgrade of the environment "to a communication channel from the more mundane role of a sink of information that it had in decoherence."[8] The observer or apparatus is understood as part of the environment now, and a property becomes more "objective" or "real" the more subsystems of the environment have recorded it. "Repeatability is key. Collectively, the environmental fragments act like the apparatuses," Zurek stresses.[9] In other words, "the redundancy of information transferred from the system to many fragments of the environment leads to the perception of objective classical reality," as Zurek says in defending his notion of reality that introduces a selection process about which information can be replicated most efficiently.[10] Musing about the other extreme, Zurek wonders, "If there is no record of an event, has it really happened?"[11]

What results is a scenario Zurek calls quantum Darwinism. He describes it as such since "clearly, there is survival of the fittest, and fitness is defined as in natural selection—through the ability to procreate."[12] According to Zurek, this survival mechanism explains why nature chooses locality over spread-out waves, combined with the characteristics of suitable environments: "Not all environments are good witnesses. However, photons excel: They do not interact with air or with each other, and so they faithfully pass on information. A small fraction of a photon environment usually reveals all an observer needs to know," Zurek explains. This is how information about position gets recorded and becomes accessible for the observer: "Objects of interest scatter air and photons, so both environments acquire information about position and favor similar localized . . . states," Zurek points out.[13] While the observer is still needed, and while "accounting for collapse goes beyond mathematics, as it involves perception" and "that is where quantum physics gets personal," as Zurek readily admits, the redundancy of copies promotes this perception to a common truth that exists, "as was the case in the classical world we once thought we inhabited . . . objectively, untouched by our curiosity and oblivious to our indirect monitoring," as Zurek writes.

Crucial to Zurek's approach is the stability and reliability of records. But when are records stable and reliable? As has been argued by Sean Carroll, records are only stable in an entropy-increasing universe, a universe expanding in time. In hypothetical universes with decreasing entropy, records would more probably originate from accidental fluctuations. In his book *From Eternity to Here*, Carroll, for example, argued that a photo of a birthday party the previous year is indeed a reliable record of the birthday party as long as entropy increases, but could be equally well or even more likely created by an accidental fluctuation of atoms making up the photo in an entropy decreasing world. To employ quantum Darwinism, then, we would have to presuppose space-time, and what's more, a space-time that is expanding just like our universe, to produce a thermodynamical arrow of time. But then, in a fundamental description, there is no coarse-graining and possibly no space-time, and it is questionable how entropy and time are defined in the first place.

Quantum Darwinism provides a convincing mechanism explaining why we live in the on-screen reality we experience day in, day out. And it suggests how we should proceed to derive space, time, and the rest of physics from an underlying quantum reality: "Start with a quantum state in Hilbert space ... Divide Hilbert space up into pieces ... Use quantum information—in particular, the amount of entanglement between different parts of the state, as measured by the mutual information—to define a 'distance' between them," as Sean Carroll has described this approach. "The claim, in its most dramatic-sounding form, is that gravity (spacetime curvature caused by energy/momentum) isn't hard to obtain in quantum mechanics—it's automatic! Or at least, the most natural thing to expect,"[14] he explained in a 2016 research paper authored by him and collaborators.[15]

But does it really work? Not according to Ovidiu "Christi" Stoica. As he points out in a 2021 paper, in every model where "only the state vector exists, and the 3D-space, a preferred basis, a preferred factorization of the Hilbert space, and everything else, emerge uniquely," a scenario Stoica refers to as "Hilbert space fundamentalism," "such emerging structures cannot be both unique and physically relevant."[16] In fact, there will be "infinitely many physically distinct structures of the exact same type" as a consequence of the symmetries of quantum mechanics.[17] In other words, "the classical level of reality cannot emerge uniquely from the minimal quantum structure alone."[18]

But if the universe doesn't fix the world we live in, who does? One possible answer is no one. There may be parallel multiverses in different bases, each consisting of many universes or Everett worlds. If so, however, it is unclear why we find ourselves in the one multiverse we seem to inhabit. As an alternative, our own role may be more important than that of unconcerned observer, just sitting there enjoying the show. Are the workings of our minds somehow related to how quantum mechanics works or how it produces the Hollywood movies we live in? Max Tegmark at least thinks so: "Consciousness is relevant to solving ... the quantum factorization problem," he wrote in his 2014 paper "Consciousness as a State of Matter."[19] As does Michael Lockwood, the late Oxford philosopher, who had already written in his 1989 book *Mind, Brain and the Quantum*, "I see the preference for a particular basis as being rooted in the nature of consciousness, rather than in the nature of the physical world in general."[20] If true, this implies that before we really can understand quantum mechanics and the universe, we first have to understand ourselves.

The I in Observer

The universe is not simply "there"; it is experienced. But how is consciousness related to space, time, and matter, and how is it linked to the foundational One that underpins the quantum universe?

"Everybody knows what consciousness is," maintains the director of the University of Wisconsin's Center for Sleep and Consciousness, Giulio Tononi. "It is what vanishes every night when we fall into dreamless sleep and reappears when we wake up or dream."[21] Yet Australian philosopher David Chalmers has dubbed the question about how consciousness arises from nonconscious matter "the hard problem." If it is hard to imagine a world without time, space, or matter, it is virtually impossible to imagine a world without consciousness. As Chalmers writes, "There is nothing that we know more intimately than conscious experience, but there is nothing that is harder to explain . . . Why should physical processing give rise to a rich inner life at all? It seems objectively unreasonable that it should, and yet it does."[22]

Even worse, in a universe that is "One," after all, the "hard problem" rises to a new level of hardness. From a monistic perspective, consciousness, spirit, or mind cannot be simply conceived of as something different that bears no intimate relation with matter, space, and time. If everything is merged into an all-encompassing One, consciousness is too and has been understood from there, just like space, time, and matter.

Is quantum physics in any meaningful way related to consciousness? While most physicists will reject such speculations, hypotheses that link quantum mechanics with consciousness have a long tradition—both in physics as in pseudoscience. Already Niels Bohr had wondered whether the apparent, nondeterministic nature of quantum physics could account for our experience of free will.[23] John von Neumann and Eugene Wigner advocated the contrary idea: rather than quantum mechanics being responsible for how we perceive ourselves, our consciousness might explain how we experience quantum mechanics. Consciousness, as the Hungarian friends and Princeton colleagues of Wheeler's argued, could be responsible for a collapse of the quantum mechanical wave function and thus make sure that the contents of our minds produce an unequivocal experience. This hypothesis, von Neumann felt, might help to solve a problem that he described as "psycho-physical parallelism": How come our conscious experience exhibits a world where objects are in well-defined locations and cats are either alive or dead but nothing in between?

The reasoning is based on what is known as the "von Neumann chain" and the "Heisenberg cut." During a measurement the subject gets entangled with the object to be measured. Next, according to Hugh Everett, the world splits, and there are several observers, one to observe each possible state of the object. But now consider an outside observer, usually called "Wigner's friend," looking at both the object and the first observer. Is it possible this second observer observes the first one plus the object in a quantum superposition? If not, why? If yes, where exactly does this series of observers observing other observers, the "von Neumann chain," end and a purported Heisenberg cut occur to define where a definite classical reality emerges?

The answer suggested by von Neumann puts consciousness in charge. As Maximilian Schlosshauer writes, "To him [von Neumann], the only sure fact was that we, as observers, always perceive definite outcomes at the conclusion of the measurement," that "only at the level of the observer an explanation of our perception of manifestly definite outcomes is actually forced out by the obvious empirical constraints."[24] So, on one hand, consciousness seems to be the only place where we definitely experience that many worlds have been collapsed into one; on the other hand, it seems to be a defining feature of consciousness that it experiences a unique reality. From that consideration it appeared obvious to assume, as von Neumann and Wigner did, that it is consciousness—as an alien element in the physical description—that collapses the wave function, a conviction Wigner only gave up on in the late 1970s after he had learned from Zeh's early work about decoherence that no collapse or Heisenberg cut is necessary to understand a quantum measurement.[25]

An alternate link between consciousness and quantum mechanics is described by Roger Penrose, the 2020 Nobel Prize winner who, together with Stephen Hawking, demystified many properties of black holes. Penrose holds that the mind itself is a quantum phenomenon and that quantum mechanics makes our minds special. Since our minds seem capable of performing tasks no computer would be able to, Penrose argued in his 1989 book *The Emperor's New Mind*, the explanation had to lie in a field beyond classical physics, and the only possible candidate would be mysterious quantum mechanics. Yet again, decoherence may act as a spoilsport. Potential neural correlates of consciousness, "while small on a biological scale, are still macroscopic and very complex objects on the typical scales considered in quantum physics," Schlosshauer points out, and even worse, they are "embedded into a macroscopic 'warm and wet' environment," justifying the expectation they should be subjected to fast and efficient decoherence.[26] Indeed, decoherence times of neurons calculated by Max Tegmark amounted to fractions of attoseconds, "much shorter than the relevant . . . timescales . . . for regular neuron firing," as Tegmark writes.[27] In view of these results, it is not easy to imagine how brain processes may preserve their genuine quantum properties.[28] What's even more important, today we can say with some authority that quantum mechanics isn't the exception, confined to the strange laws of the microcosm, but rather the rule, governing the entire universe. As such it can't be responsible for an exceptional feature such as the amazing capabilities of the mind. Moreover, with decoherence, a quantum collapse isn't needed anymore, and its apparent existence can be explained as a feature of our specific perspective, confining our view of the universe to a specific and single Everett branch.

Yet there is still another option for how quantum mechanics and consciousness may be related. Decoherence is a product of our specific perspective onto the world. Somewhere this perspective has to be determined; what self, object, and environment are has to be defined. If the apparent quantum collapse and the experience of free will, space, time, and matter are artifacts of our perspective onto the universe, and if it is unexplained how this perspective is fixed, couldn't it be possible that our perspective is determined by the way consciousness works? Can consciousness somehow—through the backdoor of the factorization problem—sneak back into the debate?

The Self of the Observer

When science writer Philip Ball discussed how Everett's "Many-Worlds Interpretation has many problems," his worries concentrated on the question of how to reconcile it with consciousness and the feeling of having a "self." "There are two (or more) versions of the observer where before there was one . . . What can it mean to say that splittings generate copies of me? In what sense are those other copies 'me'?" Ball asks.[29] He questions how Everett's interpretation can possibly account for our experience as individuals: "Consciousness relies on experience, and experience is not an instantaneous property . . . You can't 'locate' consciousness in a universe that is frantically splitting countless times every nanosecond, any more than you can fit a summer into a day," Ball writes, concluding that the many worlds interpretation "is dismantling the whole notion of selfhood. It is denying any real meaning of 'you.'"[30]

In defense of Everett, one may want to object that we don't understand consciousness in any other interpretation of quantum mechanics either. Moreover, consciousness is a phenomenon that doesn't exist on the level of individual atoms—so why should it exist on the level of other Everett branches? From all that we know, the feeling of "self" isn't fundamental; rather it is a construction based on the classical experience that emerges in the "decohered" frog perspective. In this sense the problem is similar to the moot debate about whether free will exists if physics appears to be deterministic or whether the forest exists if it consists of trees. Finally, if time and space are absent in the fundamental quantum description, why should any concept of "selfhood," obviously relying on these preconditions, exist? As Annaka Harris writes in her recent book *Conscious*, "The mystery of consciousness is related to the mystery of time: our awareness is experienced across time and cannot be separated from it."[31] Time, however, as we have seen, seems to disappear in quantum gravity. Philosopher Michael Lockwood confirms this view and adds that while "temporal flow has no place within the physicist's world view, so we must consign it to the mind"; on the other hand, it "seems to be of the very essence of consciousness and, as such, as inescapable a feature of our being as is consciousness itself."[32]

Indeed, according to the many worlds interpretation, it seems reasonable to assume that both the conscious self and a quasi-classical world (including time) emerge together. One has to keep in mind that the entangled quantum universe only *looks like* many worlds (or better, one of many worlds) from the perspective of a local observer.[33] For this reason, Zeh and philosophers such as David Albert, Barry Loewer, and Michael Lockwood have preferred to speak of "many minds" rather than "many worlds." As Jeffrey A. Barrett, philosopher at the University of California, Irvine, explains, "One might understand Everett's branches as describing the states of different minds rather than different worlds. On this sort of theory an observer's determinate experiences and beliefs are explained by the fact that he always has a determinate mental state."[34] This essentially boils down to an inversion of Ball's problem, arguing that it is exactly the notion of self that solves the problem of psychophysical parallelism and makes sure that we experience a unique reality, since "what we want is an 'interpretation' which explains how it is that we always 'see' (mistakenly so, if the many worlds interpretation is correct) macroscopic objects as not being in superpositions and never experience ourselves as in superpositions," as philosophers David Albert and Barry Loewer wrote in their "Interpretation" of the many worlds interpretation.[35]

In fact, Everett had already pondered Ball's problem. While comparing the quantum observer with a splitting amoeba, he wrote, "The question of the identity or non identity of two amoebas at a later time is somewhat vague. At any time we can consider two of them, and they will possess common memories up to a point (common parent) after which they will diverge according to their separate lives thereafter."[36] Zeh later discussed the question from the viewpoint of the observer, after the splitting had occurred: "One has to accept from experience that consciousness is realized in one of these worlds, only. This experience is similar to the one that tells us that consciousness is always realized in one person at a time," Zeh wrote in his 1967 draft of "Problems in Quantum Theory."[37] While Zeh's argument may add plausibility to why we can maintain our sense of self despite existing in many Everett worlds (and, again, seems to anticipate the recently discussed links between different locations in space-time and alternative quantum realities as in the ER=EPR conjecture), it opens a new can of worms by again relating the psychophysical parallelism to the nature of consciousness.

Unfortunately, if there is anything even harder to demystify than the origins of space, time, and matter, it is the nature of consciousness. For physicists, "a conscious person is simply food rearranged," Max Tegmark stressed in a 2014 TEDx talk.[38] After all, how could a bunch of matter, a conglomeration of particles, suddenly start to think and feel—the same kinds of particles we know well to appear entirely unconscious when they are arranged into a cucumber or an apple? Also the observation that the very same brain can be conscious at times but unconscious at others is strong evidence for the assumption that consciousness is not some new substance or principle but rather a way that matter is organized and functioning.

An interesting solution to this problem was proposed by University of Wisconsin neuroscientist Giulio Tononi. According to him, consciousness may arise as a by-product of a certain type of information processing that is sufficiently complex and highly cross-linked and may be quantified by a computable quantity known as "integrated information," or Φ: "The more integrated the system is, the more synergy it has and the more conscious it is. If individual brain regions are too isolated from one another or are interconnected at random, Φ will be low. If the organism has many neurons and is richly endowed with synaptic connections, Φ will be high," explains neuroscientist Christof Koch, who has collaborated closely with Tononi. "Basically," he summarizes, "Φ captures the quantity of consciousness."[39] Tononi himself simply writes that "consciousness is integrated information,"[40] and Max Tegmark, who studies consciousness to find a solution to the quantum factorization problem, sympathizes with this philosophy: "I think consciousness is a physical phenomenon that feels non-physical because it's just like waves and computations"; in other words, "I think that consciousness is the way information feels when it's been processed in certain complex ways," he suggested in his 2014 TEDx talk.[41] To link integrated information to quantum mechanics, Tegmark has generalized the requirements Tononi and Koch developed for neural networks to arbitrary states of matter, which include the ability to store and process information in a way that is integrated and sufficiently independent of its surroundings. This way Tegmark hopes to identify preferred quantum bases that may be more beneficial for the development of consciousness than others, which is essentially an anthropic argument. In doing so, however, Tegmark needs to rely on some preexisting difference between potentially preferred bases—otherwise none could be more beneficial than any other.

As an alternative, there exists at least one second possibility that doesn't presuppose a preferred basis of the universe. This scenario takes seriously the possibility that no preferred basis exists on a fundamental level. A preferred frame may emerge then only in the specific way or perspective from which our consciousness represents the universe.

Consciousness, in this view, comes first. As Michael Lockwood explains, "In reflecting on the relation of consciousness to the matter of the brain, philosophers have been apt to take matter for granted, assuming it is the mind rather than matter that is philosophically problematic," an attitude he traces back to our being used to "think[ing] of matter essentially along Newtonian lines." But this is a dead end, Lockwood believes: "The Newtonian conception of matter is incorrect, however, and it is high time that philosophers began properly to take on board the conception that has replaced it . . . This matter, the matter of quantum mechanics, is deeply problematic, and philosophically ill-understood."[42] As Stephen Hawking's former student Don Page, a pioneer of black hole thermodynamics and the emergent space-time program who advocates an interpretation similar to Lockwood's, adds, "The idea . . . is not that one needs to consider consciousness in order to provide a suitable interpretation of the unconscious quantum world . . . , but rather that one needs to consider consciousness precisely when one is seeking to explain properties of conscious experiences themselves."[43]

In this case the preferred basis (with everything that is entailed by it, arguably including the emergence of space, time, and matter) would be a feature of our perspective or how we perceive the universe rather than of the universe itself. Whenever in a random basis physical objects can be found that process information in a specific way, such as the integrated information processing described by Tononi, Koch, and Tegmark, consciousness would flicker into existence, and a mind would emerge that perceives the universe from its characteristic angle: "If . . . one selects a particular subsystem, and chooses some possible state of that system—not, I repeat, the state that it is 'really in,' for . . . there is no such state—then one can assign to any other subsystem a determinate quantum state, relative to the chosen state of the original subsystem," Lockwood describes, emphasizing that "a conscious subject, at any given time, is entitled to think of any other subsystem, from the whole of the rest of the universe on down, as having a determinate quantum state relative to this [i.e. 'his'] designated state." This procedure would unveil the preferred basis again as an artifact of our perspective: "What one would normally think of as the state of anything, at a given time, should really be thought of as merely its state relative to the given designated state of oneself; and this goes for the state of the universe as a whole," Lockwood explains.[44]

Meanwhile, the universe itself, in its fundamental description, remains monistic. "The universe is to be thought of as a seamless whole that evolves smoothly and deterministically in accordance with the Schrödinger equation," whereas, "to the extent that there are correlations between the subsystems, no subsystem can be thought of as being in any determinate quantum state at any given time," Lockwood writes.[45]

In our case, where the physical correlates of consciousness are neurons located in a person's brain, the person's consciousness will interact locally with the outside world, and decoherence will produce a universe where cats and pebbles, stars and planets exist in well-defined conditions and places. This, in my view, is as close as we can get to Wheeler's U, to the notion that we have by our "act of observation a part in bringing that universe itself into being."

Ego Dissolved?

Yet, if it appears now that indeed the experience of the conscious self may come first, that it is our consciousness that constitutes how quantum reality becomes classical, and that on this rock we build our world, neuroscience has some uncomfortable news in store for us. According to what many neuroscientists and philosophers believe today, the conscious self does not exist: "The self we seem to inhabit most (if not all) of the time—a localized, unchanging, solid center of consciousness—is an illusion," Annaka Harris boldly declares.[46] It boils down, to be more precise, to a mere construct originating from a process of binding and synchronization into a coherent frame.

The problems start from the moment we try to mark the boundaries of our self. In his book *The Ego Tunnel*, philosopher Thomas Metzinger describes many examples of how our sense of self may be manipulated. One of them is the rubber hand illusion experiment, performed in 1998 by psychiatrists Matthew Botvinick and Jonathan Cohen at the University of Pittsburgh. Metzinger describes this unsettling experience as follows: "The subjects observed a rubber hand lying on the desk in front of them, with their own corresponding hand concealed from their view by a screen. The visible rubber hand

and the subject's unseen hand were then synchronously stroked." In this situation, somehow the hand you see and the hand you feel get mixed up in your mind, and as a result, "suddenly, you experience the rubber hand as your own, and you feel the repeated strokes in this rubber hand. Moreover, you feel a full-blown 'virtual arm'—that is, a connection from your shoulder to the fake hand on the table in front of you."[47] Even more impressive than the rubber hand illusion is a similar experiment described by Metzinger later on in his book: a subject sees himself filmed from behind in a live broadcast while his back is being scratched, until eventually the subject locates his self in the screen in front of him. Such findings illustrate convincingly how fragile our sense of self is, so that Metzinger arrives at the conclusion that "there is no such thing as a self. Contrary to what most people believe, nobody has ever been or had a self . . . [T]o the best of our current knowledge there is no thing, no indivisible entity, that is us, neither in the brain nor in some metaphysical realm beyond this world."[48]

While, of course, the illusions described may be produced by a local process in the neural correlates of consciousness, they seem to indicate that this physical process comes first and that our experience is a construct rather than a representation, let alone the root of reality. What we have arrived at is a chicken-egg problem of quantum consciousness. The notion that a classical world is an illusion emerging from a conscious mind's perspective onto the quantum universe may help to solve the preferred basis and quantum factorization problems. Yet, according to neuroscience, the conscious self itself is an illusion, produced by a classical brain.

So maybe the emergence of "self" and a "classical world" are hopelessly entwined then, a self-reinforcing process. If we trust Douglas R. Hofstadter, author of *Gödel, Escher, Bach*, such feedback loops are indeed an essential constituent of the phenomenon of "consciousness": "It was almost as if this slippery phenomenon called 'consciousness' lifted itself up by its own bootstraps, almost as if it made itself out of nothing," Hofstadter writes in his book *I Am a Strange Loop*.[49] Such

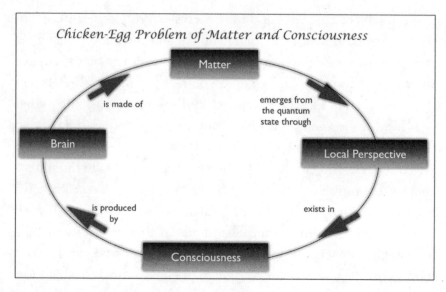

The chicken-egg problem of matter and consciousness: What comes first, matter or consciousness? Matter emerges from a local perspective onto the universe that exists in a person's consciousness, produced by the brain, which again is made of matter.

thoughts are highly speculative, of course, but if they contain some truth, it may be interesting to probe how the quantum-to-classical transition proceeds in situations where the ego seems to dissolve.

Schrödinger Cats on LSD

Somewhere around March 1, 1969, Bernard Moitessier changed the course of his ketch *Joshua* from northward to eastward. According to the rules of the Sunday Times Golden Globe Race, the first-ever round-the-world yacht race, Moitessier had circumnavigated the world single-handedly, nonstop, and unassisted in stormy waters, rounding successively the Cape of Good Hope, Cape Leeuwin, and Cape Horn. When he finally reached the calm, ice-free waters of the Atlantic Ocean, of the initial nine participants only four were left in the race. Two more would bow out later, one due to shipwreck and the other to suicide. Moitessier had excellent chances of winning the race but opted to shy away from ending his voyage, from returning to Europe and facing the media hype. Leaving his wife and his kids behind, he chose to sail on for another four months, once again around the Cape of Good Hope to Tahiti. Being alone on the ocean for six months, he was in a peculiar state of mind: "One forgets oneself, one forgets everything, seeing only the play of the boat with the sea, the play of the sea around the boat, leaving aside everything not essential to that game." University of Chicago psychologist Mihaly Csikszentmihalyi has described this state of intense absorption as "flow": "It is what the sailor holding a tight course feels when the wind whips through her hair, when the boat lunges through the waves like a colt—sails, hull, wind, and sea humming a harmony that vibrates in the sailor's veins."[50]

And, of course, flow isn't confined to sailing; it can arise in any kind of activity one feels entirely immersed in, and it is a crucial part of the creative process: "It is what a painter feels when the colors on the canvas begin to set up a magnetic tension with each other, and a new thing, a living form, takes shape in front of the astonished creator."[51] According to Csikszentmihalyi, "being in flow" corresponds to the maximal happiness possible, and moments of flow are being promoted as an integral part of a healthy and satisfying life. In more general terms, though, flow is the "man in the street's version" of what psychologists call "altered states of consciousness"—an umbrella term for states of the mind whose more extreme variants can be induced by psychoactive drugs such as LSD, psilocybin, or ayahuasca, by meditation, and by drastic incidents such as out-of-body or near-death experiences. Altered states of consciousness can go far beyond a distortion of the image of one's bodily self, culminating in the experience of a completely dissolved sense of self and time: "The legs, for example, of that chair . . . I spent several minutes—or was it several centuries?—not merely gazing at those bamboo legs, but actually being them—or . . . , to be . . . more accurate (for 'I' was not involved in the case, nor in a certain sense were 'they') being my Not-self in the Not-self which was the chair," Aldous Huxley, for example, wrote of his experience with mescalin, emphasizing that "in the final stage of egolessness there is an 'obscure knowledge' that All is in all—that All is actually each."[52] "This is as near," he suggests, "as a finite mind can ever come to 'perceiving everything that is happening everywhere in the universe.'"[53] Psychologist Marc Wittmann confirms in his book *Altered States of Consciousness* that "these states of consciousness of timelessness and eternity . . . often go hand in hand with the removal of physical and spatial boundaries and a feeling of happiness in being at one with the universe."[54] He emphasizes that "crucial to the disappearance of temporal and spatial intuition is the dissolution of the self in becoming one with the encompassing" and that "without a concept of self, time does not exist."[55]

It has been proposed that these experiences result from an inhibited information filtering usually proceeding within the thalamus as a consequence of intoxication. Metzinger agrees that the constructs we perceive as "self" and "reality" get impaired when the information processing in the brain is altered. According to Metzinger, "Conscious experience is like a tunnel. What we see and hear, or what we feel and smell and taste, is only a small fraction of what actually exists out there. Our conscious model of reality," he explains, "is a low-dimensional projection of the inconceivably richer physical reality surrounding and sustaining us."[56] In altered states of consciousness, this tunnel becomes holey and porous. UK psychologist Susan Blackmore agrees: Metzinger's "'self-model theory of subjectivity' comes close to something else I concluded—that it is not physical human beings or cats, dogs or bats that have subjective experiences but the models they create of themselves."[57] This view resonates with Chilean immunologist and neuroscientist Francisco Varela's concept that "organisms have to be understood as a mesh of virtual selves."[58] These include the "immune self," the "conscious self," and others. Is, in this case, the observing self of quantum mechanics identical to our conscious self? If not, how are the two related? According to Metzinger and Blackmore, the "conscious self" is different from the world-model-building routine inside our brains, which should be identified with the "self of the observer." But then, the fundamental outside reality is quantum mechanical until the local perspective of our self is folded in, giving rise to decoherence. Without this local perspective relying on the observing self, it remains unclear how the brain as a classical object can exist.

What's remarkable about these aspects of a psychedelic experience is that they subjectively seem to be just the opposite of what the adoption of the local frog perspective does to trigger the quantum-to-classical transition. Whereas in the frog perspective, a classical, local self and possibly even space and time emerge as a consequence of ignorance of information about the environment, conversely, both in the quantum mechanics' bird perspective and in drug-induced ecstasy, more complete knowledge seems to trigger a timeless, nonlocal experience that "all is One." Given these coincidences, one may speculate whether the local algorithm constituting the conscious self gets so strongly coupled with the environment as a consequence of hallucinogenic intoxication that it is lifted to a less local perspective and in this way is able to experience some kind of "quantum holism." Yet, whatever the relation is of the conscious self and the self of the observer, it seems to be necessary that physicists and neuroscientists confer to unravel these relations for a complete understanding of how what we perceive as reality emerges. Of course, it is far from established whether there exists any real connection between these phenomena. The apparent similarities may result from a limited number of concepts to describe nature, from a transpiration of cultural coining into the terms of scientific model building, from a structural similarity in two fundamentally different processes, but finally also from a direct relation of whatever kind between consciousness and the quantum-to-classical transition.

In 2016, I proposed an experimental setup that could probe such relations. Over the following year I developed this possibility further in collaboration with Marc Wittmann in a paper written for the Foundational Questions Institute's essay contest.[59] The basic idea is to employ a group of subjects, for example, under the influence of a psychedelic drug, such as LSD, to perform quantum measurements (such as spin-up versus spin-down) on a computer screen, while an equally prepared control group deals with an identical-looking interface connected to a classical simulation based on a random number generator. For any classical observer, the result of the measurement process would look exactly like random numbers. Thus any experience differing from seeing random results while watching quantum measurements would indicate a nonclassical perspective. As a consequence, such types of experiment may allow us to test whether the first group experiences quantum superpositions while the control group doesn't.

But is it at least in principle possible that consciousness can be entangled with anything outside the brain? Lockwood, for one, wouldn't dismiss such speculations out of hand: "I do not wish . . . wholly to prejudge the question whether experiences should be regarded as confined to the brain, or even to the body," a scenario that resonates nicely with an idea Hugh Everett's mother, Katherine Kennedy, came up with in one of her short stories:[60] "Do you believe our minds are like islands, subterraneously linked with other island minds . . . ? Do you believe, for instance, that your own individual mind is separated only so far as it believes itself separated from a kind of group mind?"[61] H. Dieter Zeh, on the other hand, is skeptical: "In the 1970s I had also tried myself to employ quantum mechanics and entanglement to learn more about the physical confinement or localization of conscious systems . . . I gave these efforts up, however, since in between a quantum mechanical measurement device and the observer's consciousness there exists a mostly quasi-classical world that also includes Tegmark's neuronal states."[62]

* * *

In the end, the thought experiment we were describing was probably too simplistic. Obviously, human consciousness doesn't interact in any meaningful way directly with a quantum system; it receives preprocessed information at the end of a chain that starts with a measurement apparatus, is transmitted by light through a medium of swirling molecules, gets fed into macroscopic sensory organs, and eventually is sent through the neural system. With technological progress, though, a more direct exposure to the quantum measurement process may become available—for example, with subjects that use neural prostheses. While any such scenarios are admittedly extremely speculative, for the time being, one can conclude that the hypothetical bird perspective onto quantum reality shares striking similarities with psychological conditions known as altered states of consciousness. At present it is far from clear where these similarities originate from, whether the experience is related to the perspectives of quantum frogs and birds, and if so, how. These questions may well open an exciting new research field about how the transition from film-roll to on-screen reality is related to the observer and how decoherence, perspective, consciousness, and the various layers of selves conspire to produce what we experience in our everyday lives.

CONCLUSION

The Unknown One

P RESUMABLY, AS SOON AS A HUMAN FOR THE FIRST time thought of herself as "I," the thought of a "you" was implied. And with that implication came the question of where "I" ceases and the "non-I," the "other," begins. This must, of course, also have brought about the thought of death: What happens when I cease to exist? The oldest myths of humanity, the roughly four-thousand-year-old Sumerian epic of Gilgamesh, and the Egyptian Book of the Dead deal with love and separation, death and immortality, and thus with unity and division, time and timelessness.

Such thoughts later evolved into monistic philosophies. Individuation and temporality came to be understood as illusionary features of an imperfect perspective, as artifacts of our perceived, on-screen reality. The true, foundational reality hidden behind our fleeting experience got conceived as a timeless, unified whole, symbolized by goddesses and gods such as Isis or Pan. The question of what unifies and what separates all beings, subjects and objects alike, appears to have absorbed humanity from its very beginning. And indeed, an obvious answer to this question is "nothing": there are no individual objects or subjects; everything and everyone is part of an integrated whole.

It is a mind-boggling fact that the same answer emerges now from our most advanced scientific theories: from quantum mechanics applied to the universe. Modern science intimately connects us to the origin of humanity—and we have seen how for thousands of years, this heritage has been a source of inspiration, as well as a burden, in the development of modern science.

Yet, after quantum mechanics has revealed its monistic foundation, and after scientists have started readily employing it in their search to make sense out of matter, space, and time, there are still open questions left: What is this "One" that we advocate as a foundation of nature? How is it related to the monistic beliefs of the ancients? What does it imply for the traditional approach to decomposing the universe into particles or strings? What does it mean for us as human beings to live in a monistic universe? And finally, can we be sure it is correct? Or could it possibly be wrong?

What Is "The One"?

At the verge of "quantum supremacy"—the moment where quantum computers outperform classical computers—the real hardware of quantum computing remains a mystery. One hundred twenty years after Max Planck's original quantum hypothesis, there is still no consensus about what quantum mechanics actually is about. If this book is right, quantum mechanics deals with "the One," an ancient philosophical concept that integrates all that possibly could happen—a three-thousand-year-old idea that has accompanied humanity from its earliest steps, through its darkest ages of science denial and religious persecution, in some of its greatest cultural accomplishments all the way to the development of modern science and quantum gravity.

But what exactly is "the One"? What is recorded on the cosmic film roll? If it stores everything that could possibly happen, are these possibilities logical possibilities, meaning that the properties of "the One" turn out to be self-evident, or a more constrained set of physical possibilities? And what is the film roll or the projector itself? Is "the One" material? Is it spirit or information? Is it math? None of the above? There is no easy resolution yet. All of these positions have their advocates.

We are in an awkward situation. On one hand, as quantum computing pioneer David Deutsch has urgently appealed, "We've got to understand this thing . . . because otherwise in every fundamental branch of physics it's as if we were planning an expedition to the moon while still thinking that the earth is flat."[1] Yet, on the other hand, as the philosopher Galen Strawson lamented in the *New York Times*, physics "tells us a great many facts about the mathematically describable structure of physical reality, facts that it expresses with numbers and equations . . . but it doesn't tell us anything at all about the intrinsic nature of the stuff that fleshes out this structure."[2] "Physics is silent . . . on this question," Strawson determines, and keeps pressing, "What is the fundamental stuff of physical reality, the stuff that is structured in the way physics reveals?"[3] At least for the moment, Strawson is perfectly right in his complaint.

One vocal champion of the hypothesis that the universe is made of information is Seth Lloyd, a pioneer of quantum information science and professor of mechanical engineering and physics at the Massachusetts Institute of Technology. Lloyd believes that the universe is properly characterized as a quantum computer:[4] "The universe is made of bits," Lloyd writes in his book *Programming the Universe*.[5] "Effectively everything is computing," Lloyd explains in an interview.[6] "In an actual computer we are accustomed socially to refer to hardware and software. But . . . this distinction is actually not as precise as what one likes to think . . . The particle is the barcode."[7]

Lloyd's way of thinking about the universe is grounded in a fundamental concept in the theory of computation. In 1936, the English mathematician Alan Turing invented the "Turing Machine," an abstract mathematical model for a universal computer. Turing had a tragic life story. As a code breaker in World War II, Turing contributed crucially to the cracking of intercepted encrypted communications between the Nazis and their co-combatants and thus to the Allies' victory. Nevertheless, when, after the war, Turing's homosexuality became known, it was prosecuted as a crime, and he was sentenced to a forced hormonal treatment that eventually drove him to commit suicide. Not until 2013, with the support of Stephen Hawking, did Queen Elizabeth II grant him a posthumous pardon.

Among his many other accomplishments, Turing contributed mightily to the foundations of information science. Just like modern computers, the Turing Machine provided an all-purpose device whose actual action was determined by the software it was running. A few years later, the German engineer Konrad Zuse—who, supported by funding from the Nazi military, had built the first programmable computer in 1941—went one step further by suggesting that the universe could be understood as "Computing Space." Later Carl Friedrich von Weizsäcker in Germany and John Wheeler in the United States developed this idea further, culminating in Wheeler's slogan "It from Bit," advocating that "every particle, every field or force, even the spacetime continuum itself—derives its function, its meaning, its very existence entirely—even if in some contexts indirectly—from the apparatus-elicited answers to yes or no questions, binary choices, bits."[8]

In fact, computer scientists use strings of zeroes and ones, so-called bits, to encode vast amounts of information that can simulate or characterize complex behavior. In quantum computing, the classical bit gets replaced by the "Q-Bit," which can be thought of as the information stored in a particle's spin pointing up or down. So far this concept of information still describes how matter is organized but not the identity of matter itself. Yet, in some cases, the Q-Bit description of a particle's spin can be generalized to determine the particle's self. As mentioned already in Chapter 6, Werner Heisenberg had pointed out in 1932 that, for example, the protons and neutrons constituting the atomic nucleus behave so similarly that they can be understood as two states of a single particle—a perfect analogy to a quantum spin or Q-Bit. Curiously, in this way the apparent material difference of proton and neutron is reduced to the information stored in the Q-Bit. An analogous argument can be made for the difference between electrons and neutrinos.[9] In the hypothetical grand unified theories developed by particle theorists since the 1970s to eventually unify all matter in the universe, this notion is taken to the extreme in that all known particles are understood now as different states of a single particle (or quantum field). In this case, the question about what exactly the hardware of the universe is becomes increasingly irrelevant, since what characterizes how the universe appears to us is determined by the information stored in this hardware rather than by the hardware itself. Just as the same USB stick can store different movies or songs and thus entirely different experiences and emotional evocations, or as computers of different makes can play the same movie and evoke the same experiences or emotional evocations irrespective of how they are built internally, the underlying hardware becomes meaningless, and what has meaning is how this hardware is organized, what information it actually stores and processes.

This is as close as it gets to Plato's book *Timaeus*, in which the philosopher develops the conception that everything we experience as outside reality is produced by informational patterns imprinted in the "midwife of being." As Lloyd emphasizes, "A quantum computer can simulate any local quantum system," meaning that indeed "the standard model and (presumably) . . . quantum gravity"—in other words, everything—"can be directly reproduced by a quantum . . . automaton."

According to Lloyd, such a picture can explain why the universe appears so complex while the actual laws of physics are apparently quite simple: "The reason is that many complex, ordered structures can be produced from short computer programs, albeit after lengthy calculations," Lloyd writes.[10] "In the beginning was the bit," Lloyd explains and goes on to flesh his idea out as follows: "As soon as the universe began, it began computing. At first, the patterns it produced were simple, comprising elementary particles and establishing the fundamental laws of physics. In time, as it processed more and more information, the universe spun out ever more intricate and complex patterns."[11] Examples include, according to Lloyd, "galaxies, stars, and planets. Life, language, human beings, society, culture—all owe their existence to the intrinsic ability of matter and energy to process information."[12] A famous example for a simple computer program giving rise to a plethora of complex structures is the Mandelbrot set, where new, aesthetic patterns are unveiled over and over again upon zooming into the fractal picture produced as an output. In a paper written in 1996, Max Tegmark took this idea to the extreme, suggesting that the universe does "in fact contain almost no information."[13] As Tegmark argues, "Decoherence together with the standard chaotic behavior of certain non-linear systems will make the universe appear extremely complex to any self-aware subsets that happen to inhabit it now, even if it was in a quite simple state shortly after the big bang."[14] In fact, if we return once more to Plato's cave, the shadows of things aren't the consequence of anything being added to the light of the sun but instead of light being screened from hitting the cave's wall, just like art carved into wood takes wood away instead of adding anything. If this is more than just a loose analogy, maybe "the One" is close to an empty canvas or a projector with no movie mounted.

But do these arguments really imply that "information comes first"? After all, information acquires its meaning only if we possess the proper software and operating system to extract this information from the hardware and process it. If we can play a recent blockbuster movie or a Leonard Cohen song from a memory stick, this doesn't make the memory stick "information." The stick is still a piece of metal, and its configuration is given by the exact states of the particles it is made of. With the help of proper software and another hardware device to play it, we can interpret this configuration as the information (i.e., the movie or song) we want to enjoy. Likewise, if we want to see how a neutrino or electron behaves, we need other quantum fields, such as force carrier or "gauge" quantum fields, to mediate their effects. Information has to be physically incorporated to become effective. Nobody—to give a drastic example—has ever been beaten to death with a Beethoven symphony, unless maybe it was scribbled on a piece of rock.

Arguably the most radical version of the idea that the universe is nothing but information has been proposed by Max Tegmark. "I will push this idea to its extreme and argue that our universe is mathematics in a well-defined sense," Tegmark wrote.[15] As he explains, "Whereas the customary terminology in physics textbooks is that the external reality is described by mathematics," he goes (at least) one step further and states that "it is mathematics."[16] Tegmark's argument is similar to Lloyd's, which is based on Turing's universal computer: "If two structures are [having a one-to-one correspondence], then there is no meaningful sense in which they are not one and the same," Tegmark writes.[17] Anything else, such as matter, mind, space, and time, is dismissed by him as "baggage." "One could argue that our universe is somehow made of stuff perfectly described by a mathematical structure, but which also has other properties that are not described by it, and cannot be described in an abstract baggage-free way," Tegmark admits, but he insists that "those additional bells and whistles that make the universe nonmathematical by definition have no observable effects whatsoever." But is this correct? Are these "bells and whistles" that make the universe nonmathematical really unobservable? One could argue that the physical realization of the mathematical structure describing the universe is what makes the universe observable in the first place, raising the concern that Tegmark's proposal confuses reality with the model describing it, making it a prime example of what philosophers call a "category mistake."

It thus appears to be more correct to think of information not as coming first but rather as a convenient way to speak about how the underlying hardware, how "the One," is organized. But there is yet another twist in this story: everything we know or experience about the world we know or experience only insofar as it exists or is represented in our conscious minds. For us, only that exists which we are conscious of, and consciousness arguably is processed information. So if everything exists exclusively in our minds, and if our minds are nothing but information, doesn't this again imply that everything is information? It appears as if we are running in circles, removing ourselves more and more from our actual experiences and observations. Just as in discussing the priority of consciousness versus matter, we run into a chicken-egg problem when we try to find out what is prior: matter or information.

But maybe this is the wrong question to ask. Maybe we shouldn't wonder whether the projector reality is either information, like that stored in an on-screen book or on a memory stick or hard disk, or

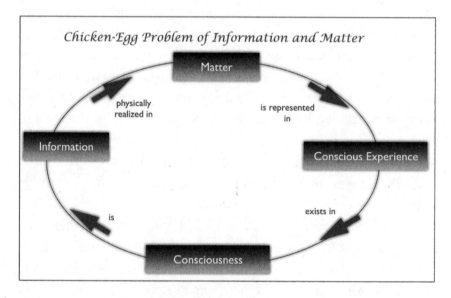

The chicken-egg problem of matter and information: What is more fundamental, matter or information? Matter is represented in conscious experience that arguably results from information processing in the brain. Information, however, is physically realized in matter.

matter, like an on-screen brick, chair, or house. Maybe we should do the converse: understand the classical, on-screen reality as information about the quantum realm; see the movie plot experienced on-screen as information about what exists in the projector room rather than understanding the film roll as information about the story that unfolds on the screen.

When I discussed these problems with H. Dieter Zeh, he wrote to me, "In this context I consider ['real'] only as 'physically real' . . . Therefore for me also e.g. the constitutional law isn't real, but only its 'realization' on paper or within a material brain."[18] When asked whether he would describe the quantum mechanical wave function as "material," though, he said no: "While I wouldn't call the wave function 'material,' I would call it 'real' . . . This is something entirely different as 'only math' or 'only information.'"[19]

As Zeh put it in "The Wave Function: It or Bit," his invited chapter for a book to celebrate John Wheeler's ninetieth birthday, "If 'it' (reality) is understood in the operationalist sense, while the wave function is regarded as 'bit' (incomplete knowledge about the outcome of potential operations), then one or the other kind of 'it' may indeed emerge 'from bit.'" Zeh admitted that for practical purposes, such a mind-set is useful: "I expect that this will remain the pragmatic language for physicists to describe their experiments for some time to come." Yet he continued to contrast this pragmatic attitude with an approach asking for the foundations of reality: "However, if 'it' is required to be described in terms of not necessarily operationally accessible but instead universally valid concepts, then the wave function remains as the only available candidate for 'it' . . . However you turn it: In the Beginning Was the Wave Function."[20]

Following Zeh, we have eventually arrived at an understanding of quantum mechanics that is diametrically opposed to Niels Bohr's view. Instead of conceiving the wave function as a tool providing information about the potential behavior of the classical objects in everyday life, this new view suggests nothing less than the contrary: classical objects, space, time, and matter have to be conceived as information

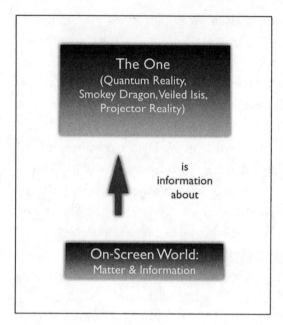

From a monistic perspective, both matter and information experienced in the daily-life, on-screen reality reveal information about the hidden reality, the fundamental "One."

about the underlying quantum reality. The behavior of classical objects allows us to constrain the space of possibilities that characterizes this fundamental reality, and the more we learn about quantum cosmology, the better we will understand what "the One" really is. Once more we can stress the parallels with Plato, who envisaged the way to truth as a strenuous ascent out of the cave, or with the metaphor of the veiled Isis, whose veil provides us with limited information about what hides beneath it.

Particles and the Universe

We started this book by realizing that if "all is One," it doesn't make sense anymore to think about the universe as composed of particles. Rather, the opposite is true: any conglomeration of particles is nothing but a specific perspective on the all-encompassing One. Such a view does nothing less than turn the quest for the foundations of physics upside down. Taking this book's argument to its logical conclusion, physics can only move forward by building on quantum cosmology instead of particles or strings. At some point—and the persisting fine-tuning problems that researchers encounter may indicate that this point has been reached—probing continually smaller distances and higher energies won't help us get closer to the foundations of physics.

This provokes the obvious question about what role particle physics can play in the future. Does it still make sense to invest billions of dollars in future particle accelerators, or is this money better invested in other branches of science? It should be stressed that particle physics will remain important. Unless it is completely understood how and why particle physics fails to uncover the essence of the universe and provide natural explanations for dark energy and the Higgs mass, the investigation of what exactly goes wrong and where the logic of "smaller is more fundamental" reaches its limits will remain an indispensable contribution to understanding "what's really going on below." Yet particle physicists will have to accept a new, more modest role in this endeavor; the field won't be able to claim to be the silver bullet to understand the cosmos anymore. Instead, particle physics will concentrate increasingly on why and where the naive reductionism of the particle approach fails.

In this new role, particle physics will remain an important pillar for foundational physics, but it will have to be supplemented by intensified efforts in research on quantum information and quantum foundations as well as cosmology. Various subdisciplines in physics, so far conceived of as essentially unrelated, will have to close ranks to unveil the foundation of nature. Which concrete experiment may play a key role in the various stages of this endeavor will have to be assessed, based on the concrete open questions in need of investigation and an honest cost-benefit analysis.

How Could They Know?

If an ancient idea really holds the key to the future of physics, how are ancient thought and modern physics related? What enabled Greek philosophers, East Asian sages, oriental mystics, and medieval thinkers to come up with ideas that are so breathtakingly close to modern physics, without having the slightest clue about the experimental progress that made it possible? If we permit speculations to run wild, the astounding parallels between what we know about quantum reality and altered states of consciousness now may provide explanations.

Maybe it is not entirely impossible that subjects experience a quantum holism in altered states of consciousness, including what has been understood as "mystical experience" from the beginning of time. Alternatively, maybe humanity somehow preserved some unconscious memory about being "one with nature" from primordial times when individuation wasn't fully developed—an arguably paradisiac state as suggested in the interpretations of the Fall of Man and discussed in the works of John Scotus Eriugena or Friedrich Schelling. Such a memory may have been fostered subsequently in pagan religions, mystery cults, and secret societies. Or the feeling of a dissolved ego that can be experienced in altered states of consciousness could have contributed to keeping this memory alive or to reviving it from time to time.

But maybe the argument makes even more sense the other way around. Maybe it is since our ancestors didn't have a clue about quantum mechanics that an ontology based on multiple, individual agents started to emerge in the first place. Maybe for our most ancient ancestors it was evident from their experience that indeed all is One, and the conception that there are many selves in the world emerged only after humans understood themselves increasingly as individuals and such a self-conception became increasingly successful in describing and predicting what happened in their daily lives when human societies gradually evolved away from nature into culture-driven modern states characterized by division of labor.

There exist, of course, more mundane explanations as to why, for millennia, humanity has pondered monistic philosophies that at their core are amazingly similar to what quantum mechanics seems to suggest. As my friend Xerxes Tata, a theoretical particle physicist at the University of Hawaii, for example, proposed in a different context, humanity may have only a limited set of concepts available to make sense of nature. If this is true, we shouldn't be surprised that similar ideas appear in quite different contexts, such as religion, theoretical physics, and abstract mathematics. A radical version of the argument above has been produced by my friend and mentor Tom Weiler, a particle theorist at Vanderbilt University: "There are only two choices. Either All is One or All isn't One. You have a fifty percent chance to be right."[21] Tom is one of the most humorous persons I have ever met, so he may have not been entirely serious about the relative chances. But whatever the concrete chances, needless to say, I don't agree. A similar argument could be made, for example, to question Albert Einstein's genius: "Either spacetime geometry is determined by the Einstein equations or it isn't." And just as Einstein's discovery of general relativity required a huge leap in abstraction from everyday experience, so does monism. For me, it remains a profound mystery how a thought so courageous as to claim the unity of the entire cosmos could ever be thought, and what's more, in the absence of any observational evidence for it. This somehow seems to suggest that it is deeply ingrained in ourselves, as conscious residents of this wonderful universe.

The One and Us

Beyond a doubt, over the next decades, we will see the realization of increasingly large quantum systems, while we ourselves will depend more and more on technical devices that will steadily employ new insights from quantum information technology. Quantum mechanics will invade our everyday lives, making them increasingly "quantum" themselves. If "the One" is the hardware of this quantum world we are about to enter, it also will become more important and meaningful for us. How will it feel for us to live in a universe that is "One" on its most foundational level? How would such a revolution in our grasp of the universe feed back into our daily lives? Can it, in some sense, make us better humans?

We live in an age of global challenges, including new pandemics, unprecedented global climate change, and a growing world population. It appears impossible that these problems can be solved by individual social groups or nations alone. Yet humanity doesn't seem to be uniting to face these challenges. Instead, societies are becoming increasingly fragmented and polarized. Can monism be of any help in this situation? If monism implies that the notion of individual agents is an illusion, that nature instead is governed by interacting networks, can such a philosophy make us less egoistic and more conscious of each other and our natural environment? Can it support us in working together instead of against each other?

Paul Harrison, the English environmentalist who founded and presides over the World Pantheist Movement, an association representing believers in the religious variety of monism, thinks so. "The pantheist 'God' is the community of all beings. It is not a He, or a She, or an It. It is a 'We,' and a we in the broadest and most inclusive sense, embracing everything from rocks and algae, through butterflies and humans, to suns and planets," Harrison writes in his book *Elements of Pantheism*.[22] Pantheism may, Harrison hopes, convince us that "this earth is the only place where we can find or make our paradise" and persuade us to behave accordingly.[23]

Such hope is not totally unsubstantiated. As we have seen, over the course of history, monism worked as a trigger for progressive ideas and human rights; it inspired the tolerant and creative intellectual climate of the Renaissance, just like the scientific revolutions since the Enlightenment and some political revolutions since the eighteenth century. Yet monism, just like science or nature in general, won't provide us with a moral compass. We know perfectly well how nature can be just as cruel and merciless as it can be beautiful, caring, and compassionate. We can equally well find examples in nature of individuals supporting each other and of strong individuals thriving and weak ones perishing. As history has proved, monistic ideas and the appeal to nature have also been abused to justify racism and social Darwinism. To avoid such perversions, we have to rely on the moral values that have emerged and stood the test of orchestrating our social relationships over the course of history.

Moreover, nature doesn't endorse any specific ethics since there is simply nothing that isn't nature. From a scientific perspective, a lush meadow isn't more natural than a highway or power plant. Genetic engineering, nuclear power, and philharmonic symphonies are part of nature, just like trees, oceans, and beetles. True, humans tend to change their environment, and that doesn't always make it more livable. But so do other biological species and even nonliving, self-organized processes, such as viruses, floods, currents, earthquakes, and volcanic activity. From all that we know, the universe doesn't care about us, our problems, or the existence of humanity in the first place. Nevertheless, scientific concepts can change the way we feel and behave, for better or worse.

It is true that monism has also been used to exclude rather than integrate, as a unique truth to which others are expected to surrender or perish. Religious fundamentalists and political extremists have exploited such a logic to turn monism first into dualism and eventually into genocidal ideologies. Likewise, monism has been abused for pseudoscientific esotericism. It is important to keep in mind that respect for nature implies that we should not do everything that is possible and, even more, that we need to accept nature as it is—that is, that one should not project one's preconceived notions onto nature. Science deniers sometimes appeal to Goethe's emotional approach to nature since it seems to allow them to identify their personal agenda as good or "natural." It does not. Science and technology are the only way to make sense out of nature and to conceive nature as it is. Only if monism can preserve an unbiased openness for diversity, an integrative perspective, can it become a philosophical guiding principle for humanity's future. That doesn't imply that it is entirely hopeless to think that monism may make us less selfish and more open and tolerant. After all, monism changes the focus from the individual to the interdependent network of individual beings. In particular the monistic spirit influencing most (if not all) religions and science alike may help to find a common ground shared by individuals of different backgrounds and beliefs, which could develop into a catalyst for religious understanding, tolerance, and peace. Such a mind-set may indeed support us in collaborating and sustaining Earth—if it could become general consent, an integral part of human culture.

But any such effort can only be successful if monism, and science in general, can be conveyed convincingly to major portions of the entire population. If a large part of humanity feels excluded, either economically or intellectually, from the insights and benefits of monism or science in general, this will spark unrest. For the better part of humanity's history, monism, though deeply engrained in our psychological heritage, remained a luxury for a small group of privileged individuals: the pharaoh and his priests in Egypt, philosophers in Athens, Alexandria, or Rome, the circle of scholars and artists around the Medici in Renaissance Florence, the Enlightenment philosophers, the scientists of the Dutch Golden Age, the members of the Royal Society in London, and the poets and thinkers in Goethe's Weimar. Again and again, history has shown that social groups that feel excluded from a philosophy claiming universal validity turn to religious fanaticism, extremist ideologies, and, in general, a dualist worldview, as exemplified in the origins of monotheistic religion and the fall of antiquity, through the rise of the fundamentalist preacher Savonarola in Renaissance Florence, to political extremism in the twentieth and twenty-first centuries. Most recently the World Wide Web, developed as a tool for particle physicists at CERN and connecting humanity around the globe more efficiently than ever before, has become an important source of knowledge and information open to everyone and at the same time a breeding ground for pseudoscience and conspiracy theories.

In order for humanity to benefit from monism, it must become more inclusive. Only if science is transparent and plausible for the mainstream of nonscientists will it help to steer humanity away from its daunting threats. Monism can be both a ray of hope and a risk in this endeavor. Only if we stand the test of this challenge, by realizing that we all are facets of an underlying unity, connecting ourselves with our immediate, ecological environment and eventually to the entire universe, may we indeed be better prepared to sustain our planet and face future crises together.

No one says it will be easy though. Embracing the universe both as a fundamental unity, as "One," and as the diversity it appears to us to contain isn't a simple task in either science or life. But it is a task that has inspired some of our most creative and noble thoughts. It is worth the try.

Can It Be Wrong?

But back to science. If monism is indeed a universal principle pervading and connecting the cosmos, we must ask a scientific question: Can it be wrong?

Of course it can! If monism is seen as a scientific concept again, this makes it vulnerable. Every scientific theory should make predictions that are confronted with experiment. And every theory can be replaced if there is experimental evidence against it, or if it persistently fails to explain important facts in its range of application—if it isn't fruitful. The main message of this book is that monism is—indeed—fruitful for science: monism follows straightforwardly from quantum mechanics taken seriously; it provides a philosophical framework for most recent work in quantum gravity and a new perspective needed to approach the fundamental problems in particle physics and cosmology. As I hope I have succeeded in pointing out, there exists solid evidence that monism is at least a promising hypothesis. As a scientist, I can't say for sure whether monism is right or wrong—nobody can. But I regard it as the best motivated and most promising candidate for a principle defining the foundations of physics we have.

The concrete relationships between matter, information, and perspective, as well as consciousness, time, and the emergence of what we perceive as our classical milieu, are far from settled. They will define the scientific and philosophical problems of the twenty-first century.

"The One," a three-thousand-year-old philosophical concept, transcending classical physics but at the same time perfectly naturalistic, maximally observer-independent, and so close to our dearest experience, is just now obtaining concrete scientific meaning. It will play a major role in this adventure. In a time when fundamental physics is facing a serious crisis and the concept of the multiverse has been argued to be "the most dangerous idea in physics," a time when the entire scientific endeavor may be at stake, when matter, space, and time are readily abandoned as fundamental elements of reality, a different perspective onto the universe that is new and old at the same time may help—in the words of German philosopher Friedrich Schelling—"to lend wings to physics once again." It is high time to claim "the One" back for science.

ACKNOWLEDGMENTS

The thoughts summarized in this book accompanied me over many years and were shaped by discussions with many fellow scientists, philosophers, authors, and friends.

Most of all I am indebted to the late H. Dieter Zeh, who generously shared and discussed his insights and research with me in an exchange of e-mails that extended over several years. I only gradually realized (and still do realize) in how many ways he was ahead of his time. Similarly important for me were the many discussions with Claus Kiefer, in which he patiently explained to me many basic and not-so-basic facts and features of quantum mechanics, decoherence, and quantum cosmology.

I also benefitted greatly from discussing various aspects covered in this book with David Albert, Nima Arkani-Hamed, Jim Baggott, Philip Ball, Laura Baudis, Adam Becker, Johannes Brachtendorf, Peter Byrne, Sean Carroll, Allen Caldwell, Claudio Calosi, David Deutsch, Sabine Ehrmann-Herfort, George Ellis, Bernd Falke, Brigitte Falkenburg, Kurt Flasch, Ken Ford, Gwen Griffith-Dickson, Erik Hoel, Sabine Hossenfelder, Erich Joos, Hans Kloft, Jean-Marc Lévy-Leblond, Andrei Linde, Bela Majorovits, Nick Mavromatos, Annica Müllenberg, George Musser, Yasunori Nomura, Thorsten Ohl, Don Page, David Parrochia, Huw Price, Yorck Ramachers, Carlo Rovelli, Josh Rosaler, Simon Saunders, Jonathan Schaffer, Maximilian Schlosshauer,

Ovidiu "Christi" Stoica, Jochen Szangolies, Peter Tallack, Xerxes Tata, Lev Vaidman, Christoph Wand, Thomas Weiler, Marc Wittmann, Michael York, Sigrid Zeh, Wojciech Zurek, and probably many others I may have forgotten. Needless to say, that doesn't imply that any of them endorses my conclusions. All mistakes are exclusively mine.

The wonderful illustrations that help to bring some of the main characters to life are based on drawings by my mother, Frigga, and I am very grateful for the kind permission of James E. Wheeler, MD, to use his father John Archibald Wheeler's famous "U."

I am also obliged to Kari Kephart, Bela Majorovits, Marc Wittmann, and Jan Zier, who read large parts of the draft manuscript and gave me valuable feedback. My mother, Frigga, supplied me with great food and a quiet place to write in times when the COVID pandemic confined everyone to his home, and my in-laws, Barbara and Tadeusz, were always there to help with work in our house and providing loving day care for our son. More than anyone else, Sara and Hemmi lovingly supported me and endured my changing moods.

This book wouldn't exist without the efforts of my agent, Giles Anderson, and of Brandon Proia, Madeline Lee, and Thomas "T. J." Kelleher and everyone else behind the scenes at Basic Books, who were essential in making it real. At times I felt lost, trying to convey how and why it is justified to conceive the universe as "One," as well as the history of the idea and the consequences for frontier research in modern physics. The book seemed a wild ride through times and places, topics and theories, including many fascinating sideshows. If I have been the least bit successful in weaving these threads into an engaging narrative, this would not have been possible without them.

FURTHER READING

About the early and not so early history of quantum mechanics, there exist (at least) two outstanding books: Manjit Kumar's *Quantum* and Jim Baggott's *The Quantum Story*. Whereas the former delivers a consistent story of the development of quantum mechanics in the first half of the twentieth century, the latter describes moments of important discoveries and includes also topics such as modern particle physics, Hawking radiation, and the Wheeler-DeWitt equation. Both are remarkably pleasant to read. This plot continues with the story of the quantum rebels who dared to question the orthodox Copenhagen interpretation. Adam Becker's recent book *What Is Real?* provides a terrific and enjoyable account of this drama. A little more scholarly but still a great read is Olival Freire Junior's *The Quantum Dissidents*. Biographies of some of the main protagonists of my book include Peter Byrne's stellar *The Many Worlds of Hugh Everett III*, which became an indispensable source of information for me, as well as John Wheeler's wonderful autobiography written with Ken Ford, *Geons, Black Holes and Quantum Foam*. More about Wheeler and his relation with Richard Feynman can be read in Paul Halpern's marvelous *The Quantum Labyrinth*. A fun read concentrating on quantum dissidents in the United States is David Kaiser's unmissable *How the Hippies Saved Physics*.

The meaning of quantum mechanics entailed as a model for reality is discussed in David Deutsch's *The Fabric of Reality* and, more recently, in Sean Carroll's *Something Deeply Hidden*, both brilliant and revealing. I also recommend Max Tegmark's illuminating articles in *Scientific American*'s *100 Years of the Quantum* (February 2001, with John Wheeler) and *Parallel Universes* (May 2003). Two great books explaining the different ways to interpret quantum mechanics (and debunking some common misconceptions) that are critical of Everett's work are Philip Ball's *Beyond Weird* and John Gribbin's *Six Impossible Things*.

A pioneering work about a timeless universe is Julian Barbour's classic *The End of Time*, and an excellent introduction to the problem of time in general is Sean Carroll's *From Eternity to Here*. For those who can read German (or Polish), I strongly recommend Claus Kiefer's stellar book *Der Quantenkosmos*. The phenomenon of entanglement and its consequences for the physics beyond space and time are the topic of George Musser's superb *Spooky Action at a Distance*, which can be accompanied by fine articles in *Quanta Magazine*, also by Musser, as well as by his colleagues Natalie Wolchover, K. C. Cole, Jennifer Ouellette, and others, who do a great job of covering the most abstract and up-to-date research in an accessible way. Excellent books that focus on the perspective of string theory or quantum loop gravity, respectively, are Brian Greene's *The Fabric of the Cosmos* and Carlo Rovelli's *Reality Is Not What It Seems*. While there is little discussion of monism in most of these books and articles, they generally agree that there is more to quantum mechanics than the probability predictions typically taught in introductory accounts and that what's missing may have important implications for cutting-edge science and our notion of reality.

The history of monism is covered most completely and most beautifully in Jan Assmann's astounding *Moses the Egyptian* and Pierre Hadot's *The Veil of Isis*. Catherine Nixey's *The Darkening Age* provides a both unsettling and captivating account of the conflict-laden relations of monism with early Christianity. The ensuing conflicts between religion and science in the early modern period are the topic of Ingrid Rowland's *Giordano Bruno: Philosopher/Heretic* and Alberto A. Martínez's *Burned Alive*; the first provides a phenomenally written account of Bruno's life and thought, whereas the latter focuses on the trials of Bruno and Galilei and the role their Pythagorean convictions played in this context. The history of the closely entwined philosophies of Platonism and Pythagoreanism is narrated in Kitty Fergusson's wonderful *Pythagoras* and the more scholarly books *Pythagoras* by Christoph Riedweg and *Pythagoras and the Pythagoreans* and *Plato and the Post-Socratic Dialogue* by Charles H. Kahn.

Breathtaking books about the rebirth of antiquity's philosophy in general and monism in particular during the Renaissance era include Stephen Greenblatt's *The Swerve*, Walter Isaacson's *Leonardo da Vinci*, and Paul Strathern's *The Medici*. And I want to recommend at least two books about the phenomenon of Romantic Science and the Second Scientific Revolution: Richard Holmes's *The Age of Wonder*, discussing the interplay of science and Romanticism mainly in England, and Andrea Wulf's fantastic *The Invention of Nature*, detailing the biography of Alexander von Humboldt, as well as his adventures and influence mainly in the Americas. While focusing mainly on historic developments, these books, each in its own way, convincingly convey the spirit that science at its best is accompanied by a unifying worldview aiming to explain nature in its entirety.

To get to grips with the discussions about beauty in physics versus anthropic reasoning and a no-bullshit data-driven approach, I recommend reading books about each stance by the respective champions, which include Frank Wilczek's *A Beautiful Question*, Leonard Susskind's *The Cosmic Landscape*, and Sabine Hossenfelder's *Lost in Math*, all highly interesting and enjoyable in their own right. Finally, if you want to read more about whether the universe is made out of matter, information, or math, Seth Lloyd's magnificent *Programming the Universe* and Max Tegmark's fabulous *Our Mathematical Universe* are probably the books you want to start with, and if you want to dig deeper into the complex relationships between consciousness and self, there are Marc Wittmann's exciting *Altered States of Consciousness*, Thomas Metzinger's amazing *The Ego Tunnel*, and Susan Blackmore's staggering *Seeing Myself*, with Annaka Harris's recent book *Conscious* providing a concise and highly entertaining introduction to the general topic.

If you want to go beyond words and are maybe a physics or STEM student or a grown-up physicist who wants to revise her old-fashioned conception of quantum mechanics, the place to start is Leonard Susskind and Art Friedman's *Quantum Mechanics: The Theoretical Minimum*. While requiring little more than high school math, the book delivers a modern and up-to-date yet highly accessible introduction to quantum mechanics that prepares for conceiving it as the physics of state vectors in configuration space rather than wave functions developing in space and time. After that, you may continue with Maximilian Schlosshauer's *Decoherence and the Quantum-to-Classical Transition*, which covers decoherence in all its facets: from experimental evidence over theoretical models to interpretational and philosophical implications. A particular gem in Schlosshauer's book is Chapter 9, where the author discusses the potential links between quantum mechanics and consciousness. From there on, how to continue depends on what your interests are: A classic in decoherence theory is *Decoherence and the Appearance of a Classical World in Quantum Theory* by Erich Joos and coauthors, including H. Dieter Zeh and Claus Kiefer. A standard reference about the problem of time in physics is H. Dieter Zeh's *The Physical Basis of the Direction of Time*. By now you are also prepared to read the collection of works by and about Hugh Everett in *The Everett Interpretation of Quantum Mechanics* (which includes Everett's original thesis) by Jeffrey Barrett and Peter Byrne; *Many Worlds? Everett, Quantum Theory and Reality* by Simon Saunders and coauthors; *The Emergent Multiverse* by David Wallace; *The Quantum Mechanics of Minds and Worlds* by Jeffrey Barrett; or the original papers by Wojciech Zurek in *Physics Today*, including "Decoherence and the Transition from Quantum to Classical" (1991) and "Quantum Darwinism, Classical Reality, and the Randomness of Quantum Jumps" (2014), as well as those by H. Dieter Zeh, of course. Zeh's papers developing many ideas that come closest to the argument I have tried to convey here are collected on his website (now accessible at its original location at the University of Heidelberg [http://www.rzuser.uni-heidelberg.de/~as3] and hosted by Claus Kiefer's group at the University of Cologne [http://www.thp.uni-koeln.de/gravitation/zeh]). Many of them are translated into German in his entertaining and provocative book *Physik ohne Realität: Tiefsinn oder Wahnsinn?*

GLOSSARY

AdS/CFT: Conjectured correspondence between a universe with a special type of gravity (AdS refers to a negative vacuum energy) and a specific quantum field theory (CFT) defined on its boundary. Also known as "Maldacena's conjecture."

Anthropic argument: Argument that we shouldn't be surprised if the universe appears fine-tuned for human life as the universe may only be one of many universes in a multiverse, and alternative universes not fine-tuned for life wouldn't have observers wondering about them.

Ant perspective: Temporal perspective describing the experience of moving through space while time passes.

Beauty (mathematical): Idea that nature adheres to a mathematical description that is simple and elegant. Often motivated by the foundational role abstract symmetries seem to play in physics.

Bird perspective: Hypothetical perspective onto the whole universe from outside, avoiding decoherence and thus allowing us experience of the universe as a quantum object.

Bit: Unit of information, quantized in a strain of zeroes and ones.

Black hole: Remnant of a heavy, burned-out star whose gravity is so strong that nothing (not even light) can escape and on whose horizon time ceases to flow.

Brahma: Monistic concept in Hinduism; compare "Tao" and "One."

Causality: Principle that nothing happens without a cause.

Classical physics: Physics before the discovery of quantum mechanics and relativity. Largely equivalent to our daily-life experience.

Complementarity: Principle in quantum mechanics that physical objects can have different characteristics, such as of a particle and a wave, at the same time.

Copenhagen interpretation: Bohr's interpretation that understood quantum mechanics as a mere tool to make probabilistic statements about classical objects.

Decoherence: Quantum effect describing how superpositions get suppressed from the perspective of a local observer when a quantum system is interacting with an unknown environment.

Deism: Philosophical position that religion should be based on rational arguments instead of authority or revelation.

Determinism: Principle that the past determines the future.

Dualism: Belief that the history of the universe can be understood as the fight of two competing principles, such as "good" versus "evil." Opposite of monism.

Emergence: Relation between a more and a less fundamental description of nature that characterizes the existence of properties and/or phenomena in the less fundamental description that do not exist in the more fundamental description.

> **Weak**: Emerging phenomena are in principle reducible to concepts in the more fundamental description, but this is usually not feasible or useful in practice.
>
> **Strong**: Emerging phenomena are in principle independent of what happens on the more fundamental level of description.

Empiricism: Philosophy advocating that knowledge comes only or primarily from experience or experimentation.

Entanglement: Property of a composed quantum system describing that the total quantum system can be in a well-known, well-defined state while the components aren't.

Entropy: Information missing to identify the microstate of a given macrostate.

Everett interpretation: See "many worlds interpretation."

Frog perspective: Perspective of a local observer onto the quantum universe, giving rise to a decohered, classical experience.

God perspective: Hypothetical perspective onto space and time from outside, depicting temporal processes as static configurations of time and space coordinates within a timeless block universe.

Hierarchy problem: Also fine-tuning or naturalness problem. Existence of strong hierarchies and fine-tuning in scientific theories.

Higgs: Particle or quantum field responsible for breaking a fundamental symmetry and creating other particles' masses.

Holism: Philosophy advocating that the whole is more than the sum of its parts.

Inflation: Theory describing an accelerated expansion of the universe that often is identified with the beginning of the universe and can in some versions of the theory produce a multiverse of universes.

Information: Quantized knowledge.

Isis: Egyptian mother goddess. Incarnation of nature and the monistic concept in ancient Egypt.

Landscape: Realm of solutions to string theory that may be realized in the multiverse spawned in some versions of cosmic inflation.

Locality: The notion that objects have a defined place and interact with other objects only at this place.

Loop quantum gravity: Candidate for a quantum theory of gravity.

Macroscopic: Large. Often described by classical physics.

Macrostate: State of a physical system defined in terms of a nonfundamental, often macroscopic description. As macrostates arise from coarse-graining over microstates, one macrostate often corresponds to many macrostates.

Maldacena's conjecture: See "AdS/CFT."

Many worlds interpretation: Interpretation of quantum mechanics that understands the quantum state as reality, giving rise to a multitude of parallel, classical realities.

Microscopic: Small. Often described by quantum physics.

Microstate: State of a physical system defined in terms of a fundamental, often microscopic description.

Monism: Philosophy advocating that all is One, and One is all. Opposite of dualism.

> **Existence monism**: Advocates that only a single One exists.
>
> **Substance monism**: Advocates that everything is made out of a single substance.
>
> **Priority monism**: Advocates that it is the whole as single One that is fundamental, while everything else derives from it (as advocated in this book).

Monotheism: Religious belief in a single God. Not to be confused with monism, as monotheism often endorses a dualist philosophy, confronting God with the devil, heaven with hell, and so on. Opposite of polytheism.

Multiverse: Notion that there exist many universes—for example, in inflation or Everett's many worlds theory.

Naturalness: Absence of strong hierarchies and fine-tuning in scientific theories.

Neith: Egyptian mother goddess, later identified with Isis.

Neoplatonism: Platonism after Plato, emphasizing the monistic trends in Platonism. Its most important representative was Plotinus.

Newtonian physics: Physics as devised by Isaac Newton. Largely identical to classical physics.

One: Monistic concept in ancient Greek philosophy, in particular in Platonism and Pythagoreanism.

Pantheism: Religious interpretation of monism, identifying the universe with God.

Platonism: Monistic philosophy prevalent in the "Academy," the school of philosophy founded by Plato.

Polytheism: Religious belief in many gods who often characterize specific aspects of a monistic conception of nature. Opposite of monotheism.

Pythagoreanism: Monistic philosophy traced back to Pythagoras, advocating the belief that nature is governed by music and math. Closely related to Platonism.

Q-Bit: Unit of quantum information, quantized in terms of spins pointing up or down.

Quantum: Energy portion of a quantum wave or excitation of a quantum field.

Quantum field: Nonlocal quantum object, such as the quantum version of the electromagnetic field.

Quantum gravity: Theory describing the gravitational field as a quantum object. So far, only hypothetical candidate theories exist.

Reductionism: Principle that a less fundamental (usually macroscopic) description can be derived from a more fundamental (usually microscopic) one.

Relativity: Theories by Einstein revolutionizing the notion of space and time.

> **Special**: Einstein's theory making time observer dependent and merging space and time into space-time.
>
> **General**: Einstein's generalization of special relativity accounting for gravity by allowing space-time to be warped.

Schrödinger equation: Equation governing the time-evolution of the quantum mechanical wave function for simple (nonrelativistic) quantum systems.

Spin: Intrinsic rotation of a particle.

String theory: Candidate for a theory of quantum gravity, describing particles as excitations of tiny strings oscillating in nine or ten space dimensions.

Superposition: Principle in quantum mechanics describing how quantum systems, instead of having defined properties, can exist in states that feature different properties with corresponding probabilities.

Symmetry: Important principle in physics describing how a physical system whose properties are changed in a specific way still adheres to the same equations.

Tao: Monistic concept in Taoism; compare "Brahma" and "One."

Wheeler-DeWitt equation: Equation describing the quantum mechanical wave function of the universe as an unchanging, timeless state.

Wheeler's U: Sketch characterizing the idea that the universe is created through its observation by a conscious observer, who herself emerges as part of the universe.

Wormhole: Shortcut connecting distant places in the universe.

NOTES

Introduction: Stargazing

1. Humboldt 1860, pp. 35–36.
2. Barnes 1987, p. 71.

1. The Hidden One

1. Thorne 2008, p. 5.
2. Feshbach, Matsui & Oleson 1988, p. 9.
3. Wheeler & Ford 1998, p. 17.
4. Ibid., p. 303.
5. Ibid., p. 104.
6. Ibid., p. 287.
7. Wheeler 1996, p. 1. See also Halpern 2017, p. 22.
8. Misner, Thorne & Zurek 2009, p. 45.
9. "I don't know how to think without pictures," Wheeler confessed in an interview. Halpern 2017, p. 22.
10. John Wheeler, interview with Ken Ford. "John Wheeler—Wheeler's Drawing of the Big U: Concept of Observer Participancy (109/130)," video posted to YouTube by Web of Stories—Life Stories of Remarkable People, October 6, 2017, https://www.youtube.com/watch?v=ttestU-obkw. Accessed December 28, 2019.
11. Ibid.
12. We will stick here to this common term, though, actually, the term "quantum reality" is somewhat misleading as it refers to the wavy reality beyond the quanta rather than to the emergent reality of the particle-like quanta themselves.
13. Letter from Einstein to Marcel Grossmann, in Kumar 2009, p. 129.
14. Baggott 2011, p. 62.
15. Heisenberg 1972, p. 59.

16. Kumar 2009, p. 200.
17. Heisenberg 1972, p. 61.
18. Kumar 2009, p. 193.
19. Ibid., p. 201.
20. Baggott 2011, p. 66.
21. Ibid., p. 65.
22. Ibid., p. 212.
23. More accurately, it is the modulus square of the wave function's amplitude that yields the probability.
24. Max Born, letter to Albert Einstein, in Baggott 2011, p. 87.
25. Ibid., p. 73.
26. Heisenberg 1972, p. 68.
27. Ibid., p. 227.
28. Bohr 1949, p. 209.
29. Kumar 2009, p. 244.
30. Ibid., p. 221.
31. Heisenberg 1958, p. 42.
32. Heisenberg 1972, p. 77.
33. Ibid., p. 63.
34. Ibid.
35. Kumar 2009, p. 238.
36. Ibid., p. 248–249.
37. Baggott 2011, p. 100.
38. Hovis & Kragh 1993.
39. Bohr 1949, pp. 210–211.
40. Kumar 2009, pp. 246, 241.
41. Ibid., p. 246.
42. Baggott 2011, p. 93.
43. This vertical version of complementarity (according to which causality and space-time are complementary) had been advocated by Bohr in his 1927 Como Lecture; the proceedings version is Bohr 1928. See also, e.g., Baggott 2011, p. 105; Kiefer 2015, p. 13.
44. Kumar 2009, p. 279.
45. Ibid., pp. 352, 251.
46. Zeh 2018, p. 7.
47. Baggott 2011, p. 94.
48. Bohr 1949.
49. Petersen 1963, p. 12: "When asked . . . [about] an underlying quantum world, Bohr would answer, 'There is no quantum world. There is only an abstract quantum physical description. It is wrong to think that the task of physics is to find out how nature is. Physics concerns what we can say about Nature.'" It has been debated, however, whether this quote really faithfully represents Bohr's philosophy. Compare,

e.g., Mermin 2004. In any case, a validated Bohr quote is "An independent reality in the ordinary physical sense can neither be ascribed to the phenomena nor to the agencies of observation" (Baggott 2011, p. 419).

50. Kumar 2009, p. 274.
51. Susskind & Friedman 2015, p. xi.
52. Albert 2019, p. 1.
53. Schlosshauer 2008b.
54. Schrödinger 1935c.
55. Heisenberg 1930, p. 65.
56. Wheeler interview with Ken Ford.
57. Wheeler & Ford 1998, p. 323.
58. Wheeler 1990, p. 5.
59. Ken Ford, e-mail to the author, April 18, 2019.
60. Wheeler & Ford 1998, p. 338.
61. Ibid., p. 354.

2. How All Is One

1. Schrödinger 1935a, p. 555.
2. Sean Carroll, Lecture 1 of "Quantum Mechanics III (Physics 125c)," Sean Carroll, April 3, 2017, https://www.preposterousuniverse.com/wp-content/uploads/125c-2017-1.pdf (accessed March 22, 2020).
3. Kiefer 2015, p. 77.
4. Gilder 2008.
5. Kumar 2009, p. 291.
6. Ibid., p. 293.
7. Isaacson 2008, p. 410.
8. Wheeler 2000.
9. Letter from Einstein to Jerome Rothstein, May 22, 1950, in Kumar 2009, p. 353.
10. Kiefer 2015, p. v.
11. Kumar 2009, p. 303.
12. Baggott 2011, p. 104.
13. Kumar 2009, p. 312.
14. Ibid., p. 341.
15. Kiefer 2015, p. 77.
16. Ibid., p. 74.
17. Ibid.
18. Schrödinger 1935a, p. 555.
19. Ibid.
20. Susskind & Friedman 2015, p. xii.
21. Weizsäcker 1971, pp. 469, 486.
22. Bohm 1951, p. 140.

23. Einstein, Letter to Robert S. Marcus, Einstein Archives Online, http://alberteinstein.info/vufind1/Record/EAR000028196 (accessed March 29, 2020).
24. Diamond 2013, p. 9.
25. Bragdon 2002, p. 18.
26. See, e.g., Lane 1990.
27. Griffith-Dickson 2005, Location 4064.
28. See, e.g., Parrinder 1970.
29. York 2003, p. 6.
30. Carter 2014, p. 182–183.
31. Pinch 2002, p. 170.
32. Assmann 1997, Location 1556.
33. Assmann 2014, p. 110.
34. Brihadaranyaka Upanishad, Chapter 5, 15, in Roebuck 2003, Location 1454.
35. Mahadevan 1957, p. 59–60.
36. Maitri Upanishad, Book 4, 2, in Roebuck 2003, Location 6252.
37. Schopenhauer 2010, p. 28.
38. Lau 1963, pp. 5–6.
39. Ibid., p. xv.
40. Huxley 1945.
41. Albert 2008, p. 50.
42. D'Espagnat 1995.
43. Zeh 2004.
44. Heisenberg 1972, p. 213.
45. Berenstain 2020, p. 113.
46. Bohr 1953, p. 388.
47. This is the English translation of original German title of Heisenberg 1972.
48. Fritjof Capra, "Heisenberg and Tagore," Fritjof Capra, July 3, 2017, https://www.fritjofcapra.net/heisenberg-and-tagore (accessed March 2, 2020).
49. Heisenberg 1972, p. 87.
50. Ibid., p. 83.
51. Ibid.
52. Pauli 1961, p. 195.
53. Jordan 1971, p. 227.
54. King James Bible, Exodus 20:4.
55. Kumar 2009, p. 352.
56. Mermin 1989, p. 9.
57. Petersen 1963, p. 11.
58. Cassidy 2010, p. 178.
59. Heisenberg 1972, p. 118.
60. Ibid.
61. Letter from Hermann to Max Jammer, in Herrmann 2019, p. 607.

62. Letter from Hermann to Heisenberg, in ibid., pp. 525–526, translated by the author.
63. Ibid.
64. Letter from Heisenberg to Hermann, in Herrmann 2019, p. 531, translated by the author.
65. Heisenberg 1982.
66. See "Urgent Call for Unity," Wikipedia, https://en.wikipedia.org/wiki/Urgent_Call_for_Unity (accessed September 8, 2021).
67. Ibid.
68. Hermann 1935.
69. "For [Hermann], quantum mechanical phenomena are not only relative to the experimental framework of observation . . . , but also relative to the . . . specific outcome of the observation. Such a notion of relativization . . . is a key feature of Everett's formulation of quantum mechanics proposed in 1957," Lumma writes. Hermann, he concludes, "forms the missing link between the early Bohr, where the classical observer interacts with the quantum system, and Everett's framework, where the observer is fully described in quantum mechanical terms." See Lumma 1999.
70. Barrett & Byrne 2012, p. 308.

3. How One Is All

1. Planck 1950, pp. 33–34.
2. Byrne 2010, p. 153.
3. Barrett & Byrne 2012, p. 197.
4. Byrne 2010, p. 91.
5. Hugh Everett III, in DeWitt & Graham 2015, p. 149.
6. Byrne 2010, p. 142.
7. Harvey Arnold, in ibid., p. 58.
8. Jammer 1974, p. 517.
9. Byrne 2010, p. 25.
10. Ibid., pp. 289–290.
11. Ibid., p. 26.
12. Ibid., p. 38.
13. Bohm 1951, pp. 583, 624, 625.
14. Byrne 2010, p. 83.
15. Ibid., p. 90.
16. Charles Misner, in ibid., p. xii.
17. Ibid., p. 103.
18. Ibid., p. 90.
19. Barrett & Byrne 2012, p. 308.
20. Byrne 2010, p. 171.

21. Ibid., p. 170.
22. Barrett & Byrne 2012, p. 308.
23. Ibid., p. 309.
24. Thorne 2008.
25. Byrne 2010, pp. 182, 132.
26. Ibid., pp. 100–101.
27. Ibid., p. 138.
28. Barrett & Byrne 2012, p. 67.
29. Byrne 2010, p. 138.
30. DeWitt 1970; Byrne 2010, p. 5.
31. Byrne 2010, p. 160.
32. Ibid., p. 332.
33. Ibid., p. 118.
34. Ibid., p. 161.
35. Wheeler & Ford 1998, p. 139.
36. Misner, in Byrne 2010, p. xii.
37. Ibid., p. 140.
38. Ibid.
39. Ibid., p. 163.
40. Ibid.
41. Ibid., p. 164.
42. Ibid.
43. Ibid., p. 166.
44. Misner, in ibid., pp. xii–xiii.
45. Ibid., p. 182.
46. Ibid., p. 175.
47. Ibid., p. 176.
48. Ibid.
49. Freire Junior 2004.
50. Byrne 2010, p. 176.
51. DeWitt & Graham 2015.
52. Byrne 2010, p. 221.
53. Freire Junior 2015, p. 140.
54. Ibid., pp. 114–115.
55. Byrne 2010, p. 168.
56. Ibid., p. 6.
57. Ibid., p. 339.
58. Ibid., p. 326.
59. Ibid., p. 250.
60. Ibid., pp. 250–251.
61. Zurek 2009, p. 181.
62. Misner, in Byrne 2010, p. xii.

63. Ibid., p. 133.
64. Ibid., p. 174.
65. See ibid., p. 321.
66. Deutsch 2010, pp. 543, 542.
67. Deutsch 1998, p. 216.
68. Barrett & Byrne 2012, p. 312.
69. Ibid.
70. Ibid.
71. Ibid., p. 313.
72. Byrne 2010, p. 140.
73. Hugh Everett III, in DeWitt & Graham 2015, p. 43; Byrne 2010, p. 152.
74. Bohm 1951, p. 584.
75. Lockwood 1989, p. 225.
76. Byrne 2010, p. 131.
77. Barrett & Byrne 2012, pp. 111, 68.
78. Byrne 2010, p. 331.
79. Freire Junior 2009, p. 282.
80. Sigrid Zeh, e-mail to the author, February 2, 2010, translated by the author.
81. Camilleri 2009, p. 300.
82. Zeh 2012a, Location 25, translated by the author.
83. Telephone conversation with Bernd Falke, December 6, 2020, recorded and translated by the author.
84. Ibid.
85. Ibid.
86. Zeh 2012a, Location 2852, translated by the author.
87. Camilleri 2009, p. 292.
88. Zeh 1993.
89. Zeh 2018.
90. Ibid.
91. Joos 1986, p. 12.
92. Schrödinger 1952a, pp. 116, 115.
93. Ibid., p. 120.
94. Zeh 2006, p. 5.
95. Zeh 2012a, Location 2801–2811, translated by the author.
96. Zeh 2006, p. 4.
97. Zeh 1967.
98. Zeh 2006, p. 5.
99. Lockwood 1989, p. 224.
100. Ornes 2019.
101. Ibid.
102. Lee 2021.
103. Schlosshauer 2008a, p. 9.

104. Ibid.
105. Ibid., p. 10.
106. Zurek 2009, p. 181.
107. Camilleri 2009, p. 292.
108. Zeh 2006, p. 2.
109. Zeh 2012a, Location 2898, translated by the author.
110. Zeh 2004.
111. Ibid.
112. Bertlmann & Zeilinger 2002, p. 12.
113. Ibid., p. 21.
114. Ibid., p. 22.
115. Ibid., p. 28.
116. Ibid., p. 23.
117. Whitaker 2012, p. vii.
118. Bertlmann & Zeilinger 2002, p. 72.
119. Whitaker 2012, p. 2.
120. Bertlmann & Zeilinger 2002, p. 199.
121. Ibid., p. 22.
122. Ibid., p. 26.
123. Ibid., p. 61.
124. Freire Junior 2004.
125. Freire Junior 2015, p. 197.
126. Ibid., p. 306.
127. Ibid.
128. Ibid.
129. Zeh 2006, p. 10.
130. Ibid.
131. Ibid.
132. Ibid.
133. Zurek 1991.
134. Max Tegmark, e-mail to the author, July 18, 2017.
135. Tegmark 2009.
136. Zeh 1967, translated by the author.
137. Ibid.
138. Ibid.
139. Byrne 2010, pp. 205–206.
140. Zeh 1967.
141. Deutsch 2010, p. 542.
142. Wallace 2012, p. 11.
143. Deutsch 2010, p. 543.
144. Wallace 2012, pp. 38, 35.

145. Zeh 2012a, Location 76, translated by the author.
146. Tegmark 2009, p. 12.
147. Zeh 1994.

4. The Struggle for One

1. Assmann 2010, p. 41.
2. Ibid., p. 43.
3. Ibid., pp. 40–41.
4. Ibid.
5. Ibid., pp. 9, 42.
6. Ibid., p. 112.
7. Assmann 1997, Locations 845–846, 75.
8. Assmann 2010, p. 4.
9. Ibid., p. 119.
10. Burnet 1963.
11. Schrödinger 2014, p. 20.
12. Barnes 1987, p. xi.
13. Taylor 2016, Location 915.
14. Ibid., Location 933.
15. Riedweg 2005, p. 85.
16. Kahn 2013, p. 97.
17. Coxon 2009, p. 54.
18. Parmenides Fragment 2, in Palmer 2019, p. 365.
19. Parmenides Fragment 6, in ibid., p. 367.
20. Ibid., p. 17.
21. Coxon 2009, p. 64.
22. Whitehead 1979, p. 39.
23. Plotinus, The Enneads 5.2.1; see MacKenna 1991, pp. 360–361.
24. Plato: Seventh Letter, 341C-E, in Jowett 2017, p. 837.
25. Albert 2008, preface, translated by the author.
26. Plato: Parmenides 139b in Jowett 2017, Kindle Location 12243.
27. Lau 1963, p. 5.
28. Weizsäcker 1971, p. 490f.
29. King James Bible, Acts 17:24–28.
30. See Halfwassen 2004, pp. 149, 152.
31. Kahn 2001, p. 100.
32. See Carabine 1995, p. 294; Albert 2008, p. 67.
33. Kahn 2001, p. 99.
34. Ibid.
35. Ibid., p. 102–103.

36. Coleman 2008, p. 29.
37. Nixey 2017, Location 204.
38. Ibid., Location 237.
39. Ibid., Locations 364, 1292, 290.
40. Bernardi 2016, p. 28.
41. "Euclid of Alexandria," in Boyer & Merzbach 1991.
42. Bernardi 2016, p. 36.
43. Watts 2017, p. 3.
44. Halfwassen 2004, p. 164–165.
45. Freely 2010, Location 1244.
46. Adamson 2007, p. 49.
47. Freely 2010, Location 882.
48. McGinnis 2010, p. 8.
49. Nicholson 1973, pp. 79–80.
50. Carabine 1995, p. 279.
51. Ibid., pp. 279–280.
52. Ibid., pp. 299–300.
53. Luibheid 1987, pp. 127–128.
54. Ibid., p. 127.
55. Ibid., pp. 188, 108.
56. Bell 1988, p. 171.
57. Luibheid 1987, p. 141.
58. Carabine 1995, p. 299.
59. Ibid., p. 279.
60. Carabine 2000, p. 6.
61. Flasch 2013, Location 2299.
62. Brennan 2002, pp. 27, 130.
63. Carabine 2000, p. 21.
64. O'Meara 1987, 848C.
65. Ibid., 956B, 650D.
66. Ibid., 637D.
67. Ibid., 750A, 652A.
68. Ibid., 721C, 652A, 476C, 724B.
69. Carabine 2000, p. 25.
70. Ibid., p. 79.
71. Jaspers 1984, pp. 362–363, translated by the author.
72. Bamford 2000, Location 806.
73. Carabine 2000, p. 19.
74. O'Meara 1987, 520A.
75. A more contemporary adoption of Plato's *Symposium* can be found in Erich Fromm's global best-selling book *The Art of Loving* (Fromm 1956).

76. MacKenna 1991, p. 347.
77. Brennan 2002, p. x.
78. Carabine 2000, p. 23.
79. See "*Caedite eos. Novit enim Dominus qui sunt eius*," Wikipedia, https://en.wikipedia.org/wiki/Caedite_eos._Novit_enim_Dominus_qui_sunt_eius (accessed September 19, 2021).
80. Albert 2011, p. 200.
81. See Pünjer 1887, p. 43.
82. Quote from his book *Summa contra Gentiles* (also known as "Book on the truth of the Catholic faith against the errors of the unbelievers").
83. "Catholic Encyclopedia (1913)/David of Dinant," Wikisource, https://en.wikisource.org/wiki/Catholic_Encyclopedia_(1913)/David_of_Dinant.
84. Davies 1994, pp. 143, 327.
85. Ibid.
86. Ibid., pp. 83, 249.
87. Stephen Greenblatt credits this event with triggering the Renaissance in his Pulitzer Prize–awarded book *The Swerve* (Greenblatt 2012).
88. Hopkins 1990, pp. 14, 39.
89. Ibid., p. 9.
90. Ibid., p. 72.
91. Ibid., p. 68.
92. Ibid., p. 74.
93. Ibid., p. 62.
94. Ibid., pp. 16–17.
95. Ibid., p. 84.
96. Ibid., p. 43.
97. Ibid.
98. Flasch 2013, Location 9332, translated by the author.
99. Kristeller 1980, p. 91.
100. Strathern 2007, Location 4473.
101. Blackwell & de Lucca 2004, p. 93.
102. Ibid., p. 101.
103. Ibid., pp. 69, 88.
104. Imerti 1964, p. 235.
105. Ibid., p. 50.
106. Ibid., p. 241.
107. Ibid., p. 238.
108. Blackwell & de Lucca 2004, p. 7.
109. Ibid., p. 91.
110. Rowland 2008, p. 79.
111. Ibid., p. 219.

112. Martínez 2018, Location 208.
113. "Et levis est cespes quid probet esse Deum"; Leibniz 1763, p. 778, translated by the author.
114. Whitman 2004, p. 93.
115. Zeh 2012a, Location 1312, translated by the author.
116. "Catholic Encyclopedia (1913)/Giordano Bruno," Wikisource, https://en.wiki source.org/wiki/Catholic_Encyclopedia_(1913)/Giordano_Bruno (accessed September 2021).
117. Ibid.
118. Forman 2011, p. 204.

5. From One to Science and Beauty

1. Giulia Giannini, "Scientific Academies," *Encyclopedia of Renaissance Philosophy*, April 6, 2020, https://doi.org/10.1007/978-3-319-02848-4_79-2.
2. Kristeller 1980, p. 101.
3. Jowett 2017, p. 1668.
4. Ibid.
5. Kristeller 1980, pp. 97, 99.
6. Ibid., p. 109.
7. Ibid.
8. Richter 1883, Location 3735.
9. Ibid., Location 304.
10. Baring 1906, Location 1700.
11. Isaacson 2017, p. 2.
12. Baring 1906, Locations 910, 984.
13. Richter 1883, Location 196.
14. Baring 1906, Location 1440.
15. Ibid., Location 962.
16. Suh 2013, Location 1436.
17. Ibid., Location 2811, 990, 1019.
18. Richter 1883, Location 101.
19. Goddu 2010, p. 221.
20. Ibid., pp. 224, 229.
21. Ibid., p. 327.
22. Freely 2014, p. 164.
23. Martínez 2018, Location 199.
24. Ferguson 2010, p. 252.
25. Martínez 2018, Location 70.
26. Kristeller 1980, p. 156.
27. Ibid., p. 157.
28. Ibid., pp. 157–158.

29. Ehrmann 1991, p. 244. Also compare James 1995, pp. 91–92: "Every reason persuades us to believe that the world is composed with harmony, both because its soul is a harmony (as Plato believed), and because the heavens are turned round their intelligences with harmony, as may be gathered from their revolutions which are proportionate to each other in velocity."
30. Cohen 1984, p. 82.
31. James 1995, p. 93.
32. Drake 1960, pp. 183–184.
33. Van Helden 1989, p. 9R.
34. Kepler in De Padova 2011, p. 80, translated by the author.
35. Gatti 1997, p. 299.
36. Martínez 2018, Location 93.
37. Livio 2002, p. 144.
38. Ibid.
39. Ferguson 2010, p. 259.
40. Livio 2002, p. 148.
41. Ferguson 2010, p. 278.
42. Richter & Scholz 1987, p. 178.
43. Hu & White 2004.
44. See Steven Nadler, "Baruch Spinoza," *Stanford Encyclopedia of Philosophy*, revised April 16, 2020, https://plato.stanford.edu/entries/spinoza (accessed September 21, 2021).
45. Spinoza 1910, p. 337.
46. Ibid., p. 189.
47. Spinoza 2017, p. 64.
48. Spinoza 1910, p. 267.
49. d'Espagnat 1995, p. 428.
50. Spinoza 1910, p. 295.
51. Israel 2001, p. 159.
52. Ibid., p. 242.
53. Jammer 1999, p. 147.
54. Ibid., p. 43.
55. Ibid., p. 129.
56. Rowen 2002, p. 218.
57. Gleick 2004, p. 3.
58. Hermann Bondi, quoted in ibid., p. 7.
59. McGuire & Rattansi 1966, p. 108.
60. Ibid.
61. Sears 1952, pp. 229–230.
62. McGuire & Rattansi 1966, p. 109.
63. Ibid., p. 126.
64. Ibid., p. 135.

65. Ibid., pp. 120, 119.
66. Ibid., p. 108.
67. Ibid., p. 119.
68. Ibid., p. 120.
69. Ibid., p. 138.
70. Ibid.
71. Jacob 1981, p. 127.
72. Champion 2003, p. 169.
73. Jacob 2019, p. 131.
74. Assmann 2014, p. 101.
75. Champion 2003, pp. 170–171.
76. Ibid., p. 170.
77. Ibid., p. 241.
78. Jacob 1970.
79. Oxenford 2008, pp. 110–111.
80. Bowring 2004, p. 49.
81. Assmann 1999, p. 38.
82. Goldstein 2019, p. 9.
83. Ibid., pp. 95–96.
84. Assmann 2005, p. 13.
85. Assmann 1997, Locations 1615–1617.
86. Adler 1998, Locations 311–313.
87. Santner 1990, p. xxvii.
88. Titled "Epicurean Confession."
89. Massimi 2017, p. 183.
90. Schelling 2013, p. 22.
91. Schelling 2004, pp. 20–21.
92. Ibid., p. 21.
93. Ibid., p. 22.
94. Ibid.
95. Massimi 2017, p. 185.
96. Schelling 2013, pp. 8–9, translated by the author.
97. Santner 1990, p. xxvii, right after Hölderlin's emphatic description of how "to be one with all."
98. Heine 1972, 2. Buch 234, translated by the author.
99. See "Lines Composed a Few Miles Above Tintern Abbey, on Revisiting the Banks of the Wye During a Tour," July 13, 1798.
100. *Prelude*, Book 5, 222, in Wordsworth 2001, p. 73.
101. Wulf 2015, pp. 128, 245.
102. Whitman 2004, p. 413.
103. Kuhn 1977, pp. 97–98.
104. Ibid.

105. Ibid., p. 96. Kuhn explicitly mentions C. F. Mohr, William Grove, Michael Faraday, and Justus von Liebig; see ibid., p. 68.
106. Ibid., p. 98.
107. Whiteley 2018, pp. 212–213.
108. Shelley 1831, p. vii.
109. Ibid.
110. Holmes 2008, p. xviii.
111. Ibid., p. 360.
112. Whiteley 2018, p. 212–213.
113. Wulf 2015, pp. 128–129.
114. *The Temple of Nature*, Canto 1, lines 15–26 (Darwin 2019, pp. 1–2).
115. Wulf 2015, p. 308.
116. Ibid., p. 304.
117. Haeckel 2016, p. 194, translated by the author.
118. Ibid., pp. 193, 9.
119. Ibid., pp. 23, 287.
120. Massimi 2017, p. 183.
121. Hermann 1970, p. 299.
122. Holmes 2008, p. 315.
123. Heine 1972, 2. Buch 234, translated by the author.
124. Wilczek 2015a, p. 323.

6. One to the Rescue

1. Giudice 2017.
2. Ibid.
3. "Das ewig Unbegreifliche an der Welt ist ihre Begreiflichkeit." Einstein 1936, p. 315, translated by Sonja Bargmann in *Ideas and Opinions* (Bonanza 1954, p. 292).
4. Ellis 2011.
5. Hossenfelder 2018.
6. Nima Arkani-Hamed, physics colloquium at Columbia University, April 29, 2013.
7. Wolchover 2013a.
8. Harari 2015, p. 24.
9. Ibid.
10. Ibid., pp. 117, 27.
11. Ibid., pp. 32–33.
12. Ibid., p. 31.
13. Schrödinger 2014, p. 69.
14. Goldstein 2019, p. 35.
15. Carroll 2016, p. 20.

16. Schaffer 2010, p. 31.
17. Ibid., p. 32.
18. Ibid., p. 33.
19. More concretely, the normalized logarithm of the number of microstates corresponding to a given macrostate is defined as entropy.
20. Susskind 2006, p. 378.
21. Ibid., p. 379.
22. Aguirre & Tegmark 2011.
23. Nomura 2011.
24. Bousso & Susskind 2012.
25. Erich Joos, e-mail to the author, February 16, 2020. Frage: "Sollte nicht statt dessen die fundamentale Physik mit der Wellenfunktion des Universum beginnen und dann durch Dekohärenz Raum, Zeit und das Standardmodell der Teilchenphysik ableiten" Erich Joos: "Ja, das ist eigentlich das Programm. Die Frage ist nur, was man noch 'reinpacken' muss. Dass man allerdings das Standardmodell 'ableiten' kann, finde ich extrem optimistisch."
26. Witten 2018.
27. Rovelli 2014.
28. Gomes 2021.
29. Tegmark 2015a.
30. Vaidman 2016.
31. Carroll & Singh 2018.
32. H. D. Zeh, e-mail to the author, April 13, 2018. "Ihr Artikel geniesst meine volle Sympathie. Er enthält sicher einige ganz neue Gesichtspunkte philosophischer Art—auch zu meiner Interpretation," translated by the author.
33. Terra Cunha, Dunningham & Vedral 2006.
34. Nima Arkani-Hamed, e-mail to the author, May 24, 2021, translated by the author.
35. Wolchover 2022.
36. Burgess 2021, p. xix.
37. See, e.g., Balasubramanian, McDermott & Van Raamsdonk 2012; Han & Akhoury 2020.
38. Nima Arkani-Hamed, "The Inevitability of Physical Laws," talk at IAS Princeton, October 26, 2012.

7. One Beyond Space and Time

1. Pais 1982, p. 152.
2. Saunders 2000.
3. Wilczek 2016.
4. Wilczek 2015b.
5. Ibid.

6. Wheeler & Ford 1998, p. 246.
7. See Baggott 2011, pp. 368–369.
8. Ibid., p. 369.
9. Ibid.
10. Wheeler & Ford 1998, p. 149.
11. Ibid., p. 248.
12. See, e.g., Misner, Thorne & Wheeler 1973, p. 1181.
13. Wheeler & Ford 1998, p. 350.
14. Baggott 2011, p. 370.
15. Ibid., p. 371.
16. Barbour 2009, p. 2.
17. Ibid.
18. Barbour 1999, Location 179.
19. Ibid., Locations 179, 218.
20. Ibid.
21. Saunders 2000.
22. Ibid.
23. Barbour 1999, p. 70.
24. Ibid., p. 264.
25. Kiefer 1994.
26. Zeh 2012a, Location 3811, translated by the author.
27. Kiefer 2009b.
28. Ibid.
29. Ibid.
30. Ibid.
31. Barbour 2009.
32. Rovelli 2009.
33. Ibid.
34. Saunders 2000.
35. Zeh 2012a, Location 4150, translated by the author.
36. Zeh 2012b, p. 205.
37. Kiefer & Zeh 1994.
38. Byrne 2010, p. 158.
39. DeWitt & Graham 2015, p. 73; Byrne 2010, p. 158.
40. Misner, Thorne & Wheeler 1973, p. 876.
41. Bekenstein 1973, p. 2333.
42. Hawking 1993, p. 65.
43. Ibid., p. 54.
44. Baggott 2011, p. 375.
45. Polchinski 2016.
46. Ibid.
47. Ibid.

48. Ibid.
49. Maldacena & Susskind 2013.
50. Ibid.
51. Ibid.
52. Ibid.
53. Ibid.
54. Susskind 2016.
55. Susskind 2016.
56. Musser 2020.
57. Verlinde 2011.
58. Ibid.
59. Ibid.
60. Ibid.
61. Ibid.
62. Ibid.
63. Cowen 2015, p. 290.
64. Van Raamsdonk 2016.
65. Ibid.
66. Van Raamsdonk 2010.
67. Ibid.
68. Ouellette 2015.
69. Cowen 2015, p. 293.
70. Ibid., p. 291.
71. "It from Qubit: Simons Collaboration on Quantum Fields, Gravity and Information," Simons Foundation, https://www.simonsfoundation.org/mathematics-physical-sciences/it-from-qubit (accessed August 2, 2021).
72. Jaksland 2020, p. 9661.
73. Wolchover 2013b.
74. Arkani-Hamed & Trnka 2014.
75. Wolchover 2013b.
76. Marletto 2021.
77. "What Is Constructor Theory," Constructor Theory, https://www.constructortheory.org/what-is-constructor-theory (accessed September 25, 2021).
78. Gefter 2021.
79. Ibid.
80. Deutsch 2010, pp. 551–552.
81. Aspen Center of Physics Colloquium Announcement, e-mail to the author, July 14, 2021.
82. Freedman & Zini 2020.
83. Donald D. Hoffman, "The Abdication of Space-Time," *Edge*, https://www.edge.org/response-detail/26563 (accessed September 25, 2021).
84. Jaksland 2020, p. 9689.

85. Susskind 2017.
86. Carroll 2019b.
87. Sean Carroll, "Space Emerging from Quantum Mechanics," Sean Carroll, July 18, 2016, https://www.preposterousuniverse.com/blog/2016/07/18/space-emerging-from-quantum-mechanics (accessed August 9, 2021).

8. The Conscious One

1. "Ich glaube, das lassen sich viele passende Bilder finden. Ich habe auch oft an Platon's Höhlengleichnis gedacht—wir sehen nur Projektionen der Realität." E. Joos, e-mail to the author, February 21, 2020, translated by the author.
2. Zurek 2009, p. 181.
3. Or into quantum system and many environments, as in Zurek's Quantum Darwinism, explained below?
4. Tegmark 2015a.
5. Bousso & Susskind 2012.
6. Ibid.
7. Zurek 2014.
8. Zurek 2009.
9. Zurek 2014.
10. Ibid.
11. Zurek 2009.
12. Ibid.
13. Zurek 2014.
14. Sean Carroll, "Space Emerging from Quantum Mechanics," Sean Carroll, July 18, 2016, https://www.preposterousuniverse.com/blog/2016/07/18/space-emerging-from-quantum-mechanics (accessed August 12, 2021).
15. Cao, Carroll & Michalakis 2017.
16. Stoica 2021.
17. Ibid.
18. Ibid.
19. Tegmark 2015a.
20. Lockwood 1989, p. 236.
21. Tononi 2008.
22. Chalmers 1995.
23. Bohr 1953, p. 389.
24. Schlosshauer 2008a, pp. 362–363.
25. Wigner 1995, p. 271.
26. Schlosshauer 2008a, p. 367.
27. Tegmark 2000.
28. It should be mentioned that these conclusions are presently debated again; see, e.g., Ouellette 2016.

29. Ball 2018b.
30. Ibid.
31. Harris 2019, p. 103.
32. Lockwood 1989, pp. 13–14.
33. Zeh has argued about this point in *John Bell's Varying Interpretations of Quantum Mechanics* as confusing "proper" and "improper mixtures," a distinction that had been first made by d'Espagnat.
34. Barrett 2003, p. 185.
35. Albert & Loewer 1988.
36. Byrne 2010, p. 138.
37. "Man muss es aber als Erfahrung hinnehmen, dass das Bewusstsein nur in jeweils einer dieser Welten realisiert ist. Diese Erfahrung ist ähnlich derjenigen, die uns sagt, dass das Bewusstsein jeweils in einer Person realisiert ist." Zeh 1967, translated by the author.
38. Max Tegmark, "Consciousness Is a Mathematical Pattern," TEDx Cambridge, 2014, http://www.tedxcambridge.com/talk/consciousness-is-a-mathematical-pattern (accessed September 25, 2021).
39. Koch 2009.
40. Tononi 2008.
41. Tegmark, "Consciousness Is a Mathematical Pattern."
42. Lockwood 1989, p. ix.
43. Page 1995.
44. Lockwood 1989, p. 228.
45. Ibid.
46. Harris 2019, p. 48.
47. Metzinger 2009, p. 3.
48. Ibid., p. 1.
49. Hofstadter 2007, Location 415.
50. Csikszentmihalyi 1990, p. 3.
51. Ibid.
52. Huxley 2004, p. 8.
53. Ibid., p. 10.
54. Wittmann 2018, p. 24.
55. Ibid., pp. 26–27.
56. Metzinger 2009, p. 6.
57. Blackmore 2017, Location 4924.
58. Varela 1995.
59. Päs 2017; Päs & Wittmann 2017.
60. Lockwood 1989, p. 16.
61. Byrne 2010, p. 21.
62. H. D. Zeh, e-mail to the author, September 18, 2016, translated by the author: "In den siebziger Jahren habe ich auch versucht, mittels Quantenmechanik

und Verschränkung mehr über die physikalische Einengung oder Lokalisierung bewusster Systeme erfahren zu können. Dabei habe ich sogar auf split-brain-Experimente verwiesen. Ich habe die Versuche aber aufgegeben, da zwischen einem qm Messapparat und dem Bewusstsein doch eine weitgehend quasi-klassische Welt liegt, wozu auch Tegmarks quasi-klassische Neuronenzustände gehören. In meinen neueren Arbeiten habe ich aber stets versucht, konsequent zwischen diesen beiden Teilen der 'Beobachtung' quantenmechanischer Systeme zu unterscheiden, wobei anscheinend bisher nur der erste Teil seriös zugänglich ist."

Conclusion: The Unknown One

1. Deutsch 2010, p. 551.
2. Galen Strawson, "Consciousness Isn't a Mystery. It's Matter," *New York Times*, May 16, 2016, https://www.nytimes.com/2016/05/16/opinion/consciousness-isnt-a-mystery-its-matter.html; Harris 2019, p. 89.
3. Strawson, "Consciousness Isn't a Mystery"; Harris 2019, p. 89.
4. Lloyd 2013.
5. Lloyd 2007, p. 3.
6. Seth Lloyd, interview with Robert Lawrence Kuhn. "Seth Lloyd—Is Information the Foundation of Reality?," video posted to YouTube by Closer to Truth, September 21, 2018, https://www.youtube.com/watch?v=a35bKt1nuBo (accessed September 12, 2021), minute 8:01.
7. Ibid., minutes 5:31 and 10:48.
8. Wheeler 1990, p. 5.
9. In technical terms, this is known as an "Isospin Doublet."
10. Lloyd 2013.
11. Lloyd 2007, prologue, Location 53.
12. Lloyd 2007, p. 3.
13. Tegmark 1996.
14. Ibid.
15. Tegmark 2008.
16. Ibid.
17. Ibid.
18. H. D. Zeh, e-mail to the author, February 19, 2016.
19. H. D. Zeh, e-mail to the author, February 11, 2016.
20. Zeh 2004.
21. Tom Weiler, private communication.
22. Harrison 2013, p. 45.
23. Ibid., p. 58.

BIBLIOGRAPHY

Adamson. Peter. 2007. *Al-Kindi*. Oxford University Press.
Adler. Jeremy. 1998. *Hölderlin: Selected Poems and Fragments*. Penguin (Kindle Edition).
Aguirre. Anthony & Tegmark. Max. 2011. *Born in an infinite universe*, Physical Review D 84, 105002. arXiv:1008.1066.
Albert. David & Loewer. Barry. 1988. *Interpreting the many worlds interpretation*. Synthese 77 (2), pp. 195–213.
Albert. David Z. 2019. *How to teach quantum mechanics*. PhiSci Preprint 15584.
Albert. Karl. 2008. *Platonismus*. WBG Academic.
Albert. Karl. 2011. *Amalrich von Bena und der mittelalterliche Pantheismus*. In: Zimmermann. Karl (ed.). *Die Auseinandersetzungen an der Pariser Universität im XIII. Jahrhundert*. Gryuter, pp. 193–212.
Arkani-Hamed. Nima & Trnka. Jaroslav. 2014. *The Amplituhedron*. Journal of High Energy Physics 30. arXiv:1312.2007.
Assmann. Jan. 1997. *Moses the Egyptian*. Harvard University Press (Kindle Edition).
Assmann. Jan. 1999. *Hen kai pan. Ralph Cudworth und die Rehabilitierung der hermetischen Tradition*. In: Neugebauer-Wölk. Monika (ed.). *Aufklärung und Esoterik*, Hamburg, pp. 38–52.
Assmann. Jan. 2005. *Schiller, Mozart und die Suche nach neuen Mysterien*. In: Bayerische Akademie der schönen Künste, Jahrbuch 19, München, pp. 13–25.
Assmann. Jan. 2010. *The Price of Monotheism*. Stanford University Press.
Assmann. Jan. 2014. *Religio Duplex*. Wiley/Polity Press (Kindle Edition).
Baggott. Jim. 2011. *The Quantum Story*. Oxford University Press.

Balasubramanian. Vijay. McDermott. Michael B. & Van Raamsdonk. Mark. 2012. *Momentum-space entanglement and renormalization in quantum field theory*. Physical Review D 86, 045014. arXiv:1108.3568.

Baldwin. Anna. 2008. *Platonism and the English Imagination*. Cambridge University Press.

Ball. Philip. 2018a. *Beyond Weird*. University of Chicago Press.

Ball. Philip. 2018b. *Why the many-worlds interpretation has many problems*, Quanta Magazine, October 18, 2018, https://www.quantamagazine.org/why-the-many-worlds-interpretation-of-quantum-mechanics-has-many-problems-20181018 (accessed August 16, 2021).

Bamford. Christopher. 2000. *John Scotus Eriugena: The Voice of the Eagle*. Lindisfarne Books (Kindle Edition).

Barbour. Julian. 1999. *The End of Time*. Oxford University Press (Kindle Edition).

Barbour. Julian. 2009. *The nature of time*. arXiv:0903.3489.

Baring. Maurice. 1906. *Leonardo da Vinci: Thoughts on Art and Life*. Merrymount Press/e-artnow (Kindle Edition).

Barnes. Jonathan. 1987. *Early Greek Philosophy*. Penguin.

Barrett. Jeffrey A. 2003. *The Quantum Mechanics of Minds and Worlds*. Oxford University Press.

Barrett. Jeffrey A. & Byrne. Peter. 2012. *The Everett Interpretation of Quantum Mechanics*. Princeton University Press.

Becker. Adam. 2018. *What Is Real?*. Basic.

Bekenstein. Jacob D. 1973. *Black holes and entropy*. Physical Review D 7, pp. 2333–2346.

Bell. John S. 1988. *Speakable und Unspeakable in Quantum Mechanics*. Cambridge University Press.

Berenstain. Nora. 2020. *Privileged-perspective realism in the quantum multiverse*. In: Glick. David. Darby. George & Marmodoro. Anna (eds.). *The Foundation of Reality*. Oxford University Press, pp. 102–122.

Bernardi. Gabriella. 2016. *The Unforgotten Sisters*. Springer.

Bertlmann. Reinhold & Zeilinger. Anton (eds.). 2002. *Quantum [Un]speakables*. Springer.

Blackmore. Susan. 2017. *Seeing Myself*. Robinson/Little, Brown Book Group (Kindle Edition).

Blackwell. Richard & de Lucca. Robert (eds.). 2004. *Giordano Bruno: Cause, Principle and Unity*. Cambridge University Press.

Bohm. David. 1951. *Quantum Theory*. Dover.

Bohr. Niels. 1928. *The quantum postulate and the recent development of atomic theory*. Nature, 121, pp. 580–590. In: Wheeler & Zurek 1983, pp. 87–126.

Bohr. Niels. 1949. *Discussions with Einstein on epistemological problems in atomic physics*. In: Schilpp 1949, pp. 199–242.

Bohr. Niels. 1953. *Physical science and the study of religion*. In: Pedersen. Johannes. *Studia Orientalia Ioanni Pedersen Septuagenario VII*. E. Munksgaard, pp. 385–390.

Bousso. Raphael & Susskind. Leonard. 2012. *The multiverse interpretation of quantum mechanics*. Physical Review D 85 045007. arXiv:1105.3796.

Bowring. Edgar A. 2004. *The Poems of Goethe*. Digireads/Neeland Media LLC (Kindle Edition).

Boyer. Carl B. & Merzbach. Uta C. 1991. *A History of Mathematics*. John Wiley & Sons.

Bragdon. Kathleen. 2002. *The Columbia Guide to American Indians of the Northeast*. Columbia University Press.

Brennan. Mary. 2002. *John Scottus Eriugena: Treatise on Divine Predestination*. University of Notre Dame Press (Kindle Edition).

Bryson. Bill. 2010. *Seeing Further*. HarperPress.

Burgess. Cliff. 2021. *Introduction to Effective Field Theory*. Cambridge University Press.

Burnet. John. 1963. *Early Greek Philosophy*. Meridian.

Byrne. Peter. 2010. *The Many Worlds of Hugh Everett III*. Oxford University Press.

Camilleri. Kristian. 2009. *A history of entanglement: Decoherence and the interpretation problem*. Studies in History and Philosophy of Modern Physics 40 (2009), pp. 290–302.

Cao. ChunJun. Carroll. Sean & Michalakis. Spyridon. 2017. *Space from Hilbert space: Recovering geometry from bulk entanglement*. Physical Review D 95, 024031. arXiv:1606.08444.

Capra. Fritjof. 1975. *The Tao of Physics*. Shambhala.

Carabine. Deirdre. 1995. *The Unknown God*. Peeters Press.

Carabine. Deirdre. 2000. *John Scottus Eriugena*. Oxford University Press.

Carroll. Sean. 2010. *From Eternity to Here*. Dutton.

Carroll. Sean. 2016. *The Big Picture*. Dutton.

Carroll. Sean. 2019a. *Something Deeply Hidden*. Dutton.

Carroll. Sean 2019b. *The hidden truth about spacetime*. New Scientist, September 14–20, 2019.

Carroll. Sean & Singh. Ashmeet. 2018. *Mad-dog Everettianism: Quantum mechanics at its most minimal*. arXiv:1801.08132.

Carter. Howard. 2014. *The Tomb of Tutankhamun, Volume 1: Search, Discovery and Clearing of the Antechamber*. Bloomsbury.

Cassidy. David C. 2010. *Beyond Uncertainty*. Bellevue Literary Press.

Chalmers. David. 1995. *Facing up to the problem of consciousness*. Journal of Consciousness Studies 2 (3), pp. 200–219.
Champion. Justin. 2003. *Republican Learning*. Manchester University Press.
Cohen. Hendrik Floris. 1984. *Quantifying Music*. Springer.
Coleman. Janet. 2008. *The Christian Platonism of Saint Augustine*. In: Baldwin 2008, pp. 27–37.
Cowen. Ron. 2015. *The quantum source of space-time*. Nature 527, pp. 290–293.
Coxon. Allan H. 2009. *The Fragments of Parmenides*. Parmenides Publishing.
Crull. Elise & Bacciagaluppi. Guido. 2016. *Grete Hermann—Between Physics and Philosophy*. Springer.
Csikszentmihalyi. Mihaly. 1990. *Flow*. HarperCollins.
D'Espagnat. Bernard. 1979. *The quantum theory and reality*. Scientific American 241, pp. 158–181.
D'Espagnat. Bernard. 1995. *Veiled Reality*. Basic.
D'Espagnat. Bernard. 1998. *Quantum theory: A pointer to an independent reality*. arXiv:quant-ph/9802046v2.
D'Espagnat. Bernard. 2009. *Quantum weirdness: "What we call 'reality' is just a state of mind."* The Guardian, March 20, 2009.
Darwin. Erasmus. 2019. *The Temple of Nature*. Sophene.
Davies. Oliver. 1994. *Meister Eckhart: Selected Writings*. Penguin.
De Padova. Thomas. 2011. *Das Weltgeheimnis*. Piper.
Deutsch. David. 1998. *The Fabric of Reality*. Penguin (Kindle Edition).
Deutsch. David. 2010. *Apart from Universes*. In: Saunders et al. 2010, pp. 542–552.
Deutsch. David & Marletto. Chiara. 2014. *Constructor theory of information*. arXiv:1405.5563.
DeWitt. Bryce. 1970. *Quantum mechanics and reality*. Physics Today 23 (9), September 1, 1970, p. 30.
DeWitt. Bryce & Graham. Neill. 2015. *The Many-Worlds Interpretation of Quantum Mechanics*. Princeton University Press.
Diamond. Jared. 2013. *The World until Yesterday*. Penguin (Kindle Edition).
Dickie. John. 2020. *The Craft*. Hodder & Stoughton.
Dillon. John. 1991. *Plotinus: The Enneads*. Penguin.
Drake. Stillman. 1960. *Galileo Galilei: The Assayer*. University of Pennsylvania Press.
Dürr. Hans-Peter. 1990. *"Physik und Transzendenz."* Knaur.
Ehrmann. Sabine. 1991. *Marsilio Ficino und sein Einfluß auf die Musiktheorie*. Archiv für Musikwissenschaft H. 3., pp. 234–249.

Einstein. Albert. 1936. *Physik und Realität*. Journal of the Franklin Institute 221 (3), pp. 313–347.

Einstein. Albert. Podolsky. Boris & Rosen. Nathan. 1935. *Can quantum-mechanical description of physical reality be considered complete?*. Physical Review 47: 777.

Ellis. George F. R. 2011. *Why the Multiverse May Be the Most Dangerous Idea in Physics*. Originally published as *Does the Multiverse Really Exist?* Scientific American 305 (2) (August).

Emerson. Ralph Waldo. 2003. *Nature and Selected Essays*. Penguin.

Everett. Hugh. *Relative state formulation of quantum mechanics*. Reviews of Modern Physics 29 (3), pp. 454–462.

Ferguson. Kitty. 2010. *Pythagoras*. Icon Books (Kindle Edition).

Feshbach. Herman. Matsui. Tetsuo & Oleson. Alexandra. 1988. *Niels Bohr: Physics and the World*. Routledge.

Flasch. Kurt. 2004. *Nikolaus von Kues in seiner Zeit*. Reclam.

Flasch. Kurt. 2007. *Nikolaus Cusanus*. C. H. Beck.

Flasch. Kurt. 2013. *Das philosophischen Denken im Mittelalter*. Reclam (Kindle Edition).

Flasch. Kurt. 2015. *Meister Eckhart: Philosopher of Christianity*. Yale University Press.

Forman. Paul. 2011. *Weimar Culture, Causality and Quantum Theory*. In: Carson. Cathryn. Kojevnikov. Alexei & Trischler. Helmuth (eds.). *Weimar Culture and Quantum Mechanics*. World Scientific, pp. 203–119.

Freedman. Michael & Zini. Modjtaba Shokrian. 2020. *The universe from a single particle*. arXiv:2011.05917.

Freely. John. 2010. *Aladdin's Lamp*. Vintage.

Freely. John. 2014. *Celestial Revolutionary*. I. B. Tauris.

Freire Junior. Olival. 2004. *The historical roots of "foundations of quantum physics" as a field of research (1950–1970)*. Foundations of Physics 34 (11).

Freire Junior. Olival. 2009. *Quantum dissidents: Research on the foundations of quantum theory circa 1970*. Studies in History and Philosophy of Modern Physics 40, pp. 280–289.

Freire Junior. Olival. 2015. *The Quantum Dissidents*. Springer (Kindle Edition).

Fromm. Erich. 1956. *The Art of Loving*. Harper & Row.

Gatti. Hilary. 1997. *Giordano Bruno's Ash Wednesday Supper and Galileo's Dialogue of the Two Major World Systems*. Bruniana & Campanelliana 3 (2), pp. 283–300.

Gatti. Hilary. 1999. *Giordano Bruno and Renaissance Science*. Cornell University Press.

Gefter. Amanda. 2021. *How to rewrite the laws of physics in the language of impossibility*. Quanta Magazine, April 29, 2021. https://www.quantamagazine.org/with-constructor-theory-chiara-marletto-invokes-the-impossible-20210429 (accessed September 25, 2021).

Gilder. Louisa. 2008. *The Age of Entanglement*. Alfred A. Knopf.

Giudice. Gian. 2017. *The dawn of the post-naturalness era*. arXiv:1710.07663 [physics.hist-ph].

Gleick. James. 2004. *Isaac Newton*. Harper Perennial.

Goddu. André. 2010. *Copernicus and the Aristotelian Tradition*. Brill.

Goldstein. Jürgen. 2019. *Georg Forster*. University of Chicago Press.

Gomes. Henrique. 2021. *Holism as the empirical significance of symmetries*. European Journal for Philosophy of Science 11 (3), p. 87. arXiv:1910.05330.

Greenblatt. Stephen. 2012. *The Swerve: How the Renaissance Began*. Vintage.

Gribbin. John. 2005. *The Fellowship*. Allen Lane.

Gribbin. John. 2012. *Erwin Schrödinger and the Quantum Revolution*. Bantam (Kindle Edition).

Gribbin. John. 2019. *Six Impossible Things*. Icon (Kindle Edition).

Griffith-Dickson. Gwen. 2005. *The Philosophy of Religion*. SCM Press (Kindle Edition).

Hadot. Pierre. 2006. *The Veil of Isis*. Harvard University Press.

Haeckel. Ernst. 2016. *Die Welträtsel*. Zenodot.

Halfwassen. Jens. 2004. *Plotin und der Neuplatonismus*. C. H. Beck.

Halliwell. Jonathan J. Pérez-Mercader. Juan & Zurek. Wojciech H. 1994. *Physical Origins of Time Asymmetry*. Cambridge University Press.

Halpern. Paul. 2017. *The Quantum Labyrinth*. Basic.

Han. Bingzheng & Akhoury. Ratindranath. 2020. *Entanglement, renormalization and effective field theories*. arXiv:2011.05380.

Harari. Yuval Noah. 2015. *Sapiens*. Vintage (Kindle Edition).

Harris. Annaka. 2019. *Conscious*. HarperCollins (Kindle Edition).

Harrison. Paul. 2013. *Elements of Pantheism*. Element Books (Kindle Edition).

Hawking. Stephen. 1993. *Hawking on the Big Bang and Black Holes*. World Scientific.

Heine. Heinrich. 1972. *Zur Geschichte der Religion und Philosophie in Deutschland*. In: Heine. Heinrich (ed.). *Werke und Briefe in zehn Bänden*. Band 5. Aufbau-Verlag, pp. 216–257.

Heisenberg. Elisabeth. 1982. *Das politische Leben eines Unpolitischen*. Piper.

Heisenberg. Werner. 1930. *The Physical Principles of the Quantum Theory*. University of Chicago Press 1930, Dover 1949.

Heisenberg. Werner. 1958. *Physics and Philosophy*. Harper & Brothers.

Heisenberg. Werner. 1972. *Physics and Beyond*. Harper & Row.
Hermann. Armin. 1970. *Der Kraftbegriff bei Michael Faraday und seine historische Wurzel*. Physikalische Blätter 26 (7).
Hermann. Grete. 1935. *Die naturphilosophischen Grundlagen der Quantenmechanik*. Die Naturwissenschaften 23 (42), pp. 718–721.
Herrmann. Kay. 2019. *Grete Henry-Hermann: Philosophie—Mathematik—Quantenmechanik*. Springer.
Heuser-Keßler. Marie-Luise. 1992. *Schelling's Concept of Self-Organization*. In: Friedrich. Rudolf & Wunderlin. Arne (eds.). *Evolution of Dynamical Structures in Complex Systems*. Springer.
Hofstadter. Douglas R. 2007. *I Am a Strange Loop*. Basic (Kindle Edition).
Holmes. Richard. 2008. *The Age of Wonder*. Harper Press.
Hopkins. Jasper. 1990. *Nicholas of Cusa: On Learned Ignorance*. Arthur J. Banning Press.
Hornung. Erik. 2005. *Der Eine und die Vielen*. WBG.
Hossenfelder. Sabine. 2018. *Lost in Math*. Basic.
Hovis. R. Corby & Kragh. Helge. 1993. *P.A.M. Dirac and the beauty of physics*. Scientific American 268 (May).
Hu. Wayne & White. Martin. 2004. *The cosmic symphony*. Scientific American 290 (February), pp. 46–55.
Humboldt. Alexander von. 1860. *Letters of Alexander von Humboldt to Varrnhagen von Ense: From 1827 to 1858*. Rudd & Carleton.
Huxley. Aldous. 1945. *The Perennial Philosophy*. Harper.
Huxley. Aldous. 2004. *The Doors of Perception*. Vintage (Kindle Edition).
Iliffe. Rob. 2017. *Priest of Nature*. Oxford University Press.
Imerti. Arthur D. 1964. *Giordano Bruno: Expulsion of the Triumphant Beast*. Rutgers University Press.
Isaacson. Walter. 2008. *Einstein*. Pocket Books.
Isaacson. Walter. 2017. *Leonardo da Vinci*. Simon & Schuster (Kindle Edition).
Ismael. Jennan & Schaffer. Jonathan. 2020. *Quantum holism: Nonseparability as common ground*. Synthese 197, pp. 4131–4160. https://doi.org/10.1007/s11229-016-1201-2 (accessed September 2, 2021).
Israel. Jonathan I. 2001. *Radical Enlightenment*. Oxford University Press.
Jacob. Margaret C. 1970 *An unpublished record of a Masonic lodge in England: 1710*. Zeitschrift für Religions- und Geistesgeschichte 22 (2), pp. 168–171.
Jacob. Margaret C. 1981. *Radical Enlightenment*. George Allen & Unwin.
Jacob. Margarte C. 2019. *The Secular Enlightenment*. Princeton University Press.
Jaksland. Rasmus. 2020. *Entanglement as the world-making relation: Distance from entanglement*. Synthese 198, pp. 9661–9693.

James. Jamie. 1995. *The Music of the Spheres*. Abacus.
Jammer. Max. 1974. *The Philosophy of Quantum Mechanics*. John Wiley & Sons.
Jammer. Max. 1999. *Einstein and Religion*. Princeton University Press.
Jaspers. Karl. 1984. *Der philosophische Glaube*. Piper.
Joos. Erich. 1986. Quantum theory and the appearance of the classical world. In: Greenberger. Daniel M. (ed.). *New Techniques and Ideas in Quantum Measurement Theory*, pp. 6–13. New York Academy of Sciences.
Jordan. Pascual. 1971. *Die weltanschauliche Bedeutung der modernen Physik*, in Dürr 1990.
Jowett. Benjamin. 2017. *Plato: The Complete Works*. Olymp Classics.
Kahn. Charles H. 2001. *Pythagoras and the Pythagoreans*. Hackett Publishing.
Kahn. Charles H. 2013. *Plato and the Post-Socratic Dialogue*. Cambridge University Press.
Kaiser. David. 2012. *How the Hippies Saved Physics*. W. W. Norton.
Kiefer. Claus. 1994. *Semiclassical gravity and the problem of time*. arXiv:gr-qc/9405039.
Kiefer. Claus. 2009a. *Der Quantenkosmos*. Fischer.
Kiefer. Claus. 2009b. *Does time exist in quantum cosmology?*. arXiv:0909.3767.
Kiefer. Claus. 2015. *Albert Einstein, Boris Podolsky, Nathan Rosen*. Springer.
Kiefer. Claus & Zeh. H. Dieter. 1994. *Arrow of time in a recollapsing universe*. arXiv:gr-qc/9402036v2.
Koch. Christof. 2009. A *"complex" theory of consciousness*. Scientific American MIND, July 1. https://www.scientificamerican.com/article/a-theory-of-consciousness (accessed August 24, 2021).
Kristeller. Paul Oskar. 1964. *Eight Philosophers of the Italian Renaissance*. Stanford University Press.
Kristeller. Paul Oskar. 1980. *Renaissance Thought and the Arts*. Princeton University Press.
Kuhn. Thomas S. 1977. *The Essential Tension*. University of Chicago Press.
Kumar. Manjit. 2009. *Quantum: Einstein, Bohr and the Great Debate About the Nature of Reality*. Icon.
Lane. Beldon C. 1990. *The breath of God: A primer in Pacific/Asian theology*. Christian Century, September 19–26, pp. 833–838.
Lau. Darell. 1963. *Lao Tzu: Tao Te Ching*. Penguin.
Lee. Kai Sheng. et al. 2021. *Entanglement between superconducting qubits and a tardigrade*. arXiv:2112.07978.
Leibniz. Gottfried Wilhelm. 1763. *Theodicee, Band II*. Breitkopf & Sohn.
Livio. Mario. 2002. *The Golden Ratio*. Broadway Books.
Lloyd. Seth. 2007. *Programming the Universe*. Vintage (Kindle Edition).
Lloyd. Seth. 2013. *The universe as a quantum computer*. arXiv:1312.455.

Lockwood. Michael. 1989. *Mind, Brain and the Quantum.* Basil Blackwell.

Luibheid. Colm. 1987. *Pseudo-Dionysius—the Complete Works.* Paulist Press.

Lumma. Dirk. 1999. *The Foundations of Quantum Mechanics in the Philosophy of Nature by Grete Hermann.* Harvard Review of Philosophy 7, pp. 35–44.

MacKenna. Stephen. 1991. *Plotinus: The Enneads.* Penguin.

Mahadevan. Telliyavaram. 1957. *The Upanishads.* In: Radhakrishnan. Sarvepalli (ed.). *History of Philosophy Eastern and Western.* George Allen.

Maldacena. Juan & Susskind. Leonard. 2013. *Cool horizons for entangled black holes.* Fortschritte der Physik 61, pp. 781–811. arXiv:1306.0533.

Marletto. Chiara. 2021. *The Science of Can and Can't.* Viking.

Martínez. Alberto A. 2018. *Burned Alive: Giordano Bruno, Galileo and the Inquisition.* Reaktion.

Massimi. Michaela. 2017. *Philosophy and the Chemical Revolution After Kant.* In: Ameriks. Karl (ed.). *The Cambridge Companion to German Idealism.* 2nd edition. Cambridge University Press., pp. 182–204.

McGinnis. Jon. 2010. *Avicenna.* Oxford University Press.

McGuire. James E. & Rattansi. Piyo M. 1966. *Newton and the pipes of Pan.* Notes and Records of the Royal Society 21 (2), pp.108–143.

Mermin. N. David. 1989. *What's wrong with this pillow?* Physics Today 42 (4), pp. 9–11.

Mermin. N. David. 2004. *What's wrong with this quantum world?* Physics Today 57 (2), p. 10.

Metzinger. Thomas. 2009. *The Ego Tunnel.* Basic.

Misner. Carl. Thorne. Kip & Wheeler. John. 1973. *Gravitation.* W. H. Freeman & Company.

Misner. Carl. Thorne. Kip & Zurek. Wojciech H. 2009. *John Wheeler, relativity and quantum information.* Physics Today 62 (4), pp. 40–46.

Moitessier. Bernard. 1995. *The Long Way.* Sheridan House.

Moore. Walter J. 1989. *Schrödinger: Life and Thought.* Reissue 2015. Cambridge University Press.

Musser. George. 2015. *Spooky Action at a Distance.* Farrar, Straus and Giroux.

Musser. George. 2020. *The most famous paradox in physics nears its end.* Quanta Magazine, October 29. https://www.quantamagazine.org/the-black-hole-information-paradox-comes-to-an-end-20201029 (accessed August 8, 2021).

Nadler. Steven. 2011. *A Book Forged in Hell.* Princeton University Press.

Nicholson. Reynold A. 1973. *Rumi: Divani Shamsi Tabriz.* Rainbow Bridge.

Nixey. Catherine. 2017. *The Darkening Age.* Macmillan (Kindle Edition).

Nomura. Yasunori. 2011. *Physical theories, eternal inflation, and quantum universe.* Journal of High Energy Physics 63. arXiv:1104.2324.

O'Meara. John. 1987. *Eriugena: Periphyseon*. Dumbarton Oaks.

Ornes. Stephen. 2019. *News feature: Quantum effects enter the macroworld*. Proceedings of the National Academy of Sciences 116 (45) (November 5), pp. 22413–22417.

Ouellette. Jennifer. 2015. *How quantum pairs stitch space-time*. Quanta Magazine, April 28, 2015. https://www.quantamagazine.org/tensor-networks-and-entanglement-20150428 (accessed April 3, 2022).

Ouellette. Jennifer. 2016. *A new spin on the quantum brain*. Quanta Magazine, November 2, 2016. https://www.quantamagazine.org/a-new-spin-on-the-quantum-brain-20161102 (accessed April 3, 2022).

Oxenford. John. 2008. *Johann Wolfgang von Goethe—Autobiography*. Floating Press.

Page. Don N. 1995. *Sensible quantum mechanics: Are only perceptions probabilistic?* arXiv:quant-ph/9506010.

Pais. Abraham. 1982. *"Subtle Is the Lord . . . ": The Science and Life of Albert Einstein*. Oxford University Press.

Palmer. John. 2009. *Parmenides and Presocratic Philosophy*. Oxford University Press.

Palmer. John. 2019. *Parmenides*. In: Zalta. Edward N. (ed.). *The Stanford Encyclopedia of Philosophy*. https://plato.stanford.edu/archives/fall2019/entries/parmenides (accessed March 11, 2020).

Parrinder. Edward Geoffrey. 1970. *Monotheism and Pantheism in Africa*. Journal of Religion in Africa 3 (Fasc. 1), pp. 81–88.

Päs. Heinrich. 2017. *Can the many-worlds-interpretation be probed in psychology?* International Journal of Quantum Foundations 3 (1). arXiv:1609.04878.

Päs. Heinrich & Wittmann. Marc. 2017. *How to set goals in a timeless quantum universe*. FQXi. https://fqxi.org/community/forum/topic/2882 (accessed August 25, 2021).

Pauli. Wolfgang. 1961. *Die Wissenschaft und das abendländische Denken*, in Dürr 1990.

Petersen. Aage. 1963. *The philosophy of Niels Bohr*. Bulletin of the Atomic Scientists 19 (7), pp. 8–14.

Pinch. Geraldine. 2002. *Handbook of Egyptian Mythology*. ABC-Clio.

Planck. Max. 1950. *Scientific Autobiography and Other Papers*. Williams & Norgate.

Polchinski. Joseph. 2016. *The black hole information problem*. arXiv:1609.04036.

Pünjer. Bernhard. 1887. *History of the Christian Philosophy of Religion from the Reformation to Kant*. T. & T. Clark.

Rattansi. Piyo M. 1968. *The intellectual origins of the Royal Society*. Notes and Records of the Royal Society 23 (2), pp.129–143.

Richter. Jean Paul. 1883. *The Notebooks of Leonardo da Vinci*. Public Domain (Kindle Edition).

Richter. Peter H. & Scholz. Hans-Joachim. 1987. *Der goldene Schnitt in der Natur—harmonische Proportionen und die Evolution*. In: Küppers. Bern-Olaf (ed.). 1987. *Ordnung aus dem Chaos*, Piper.

Riedweg. Christoph. 2005. *Pythagoras*. Cornell University Press.

Roebuck. Valerie. 2003. *The Upanishads*. Penguin.

Roeck. Bernd. 2019. *Der Morgen der Welt*. C. H. Beck.

Rovelli. Carlo. 2009. *Forget time*. arXiv:0903.3832.

Rovelli. Carlo. 2014. *Why gauge?* Foundations of Physics 44, pp. 91–104. arXiv:1308.5599.

Rowen. Herbert H. 2002. *John De Witt*. Cambridge University Press.

Rowland. Ingrid. 2008. *Giordano Bruno: Philosopher/Heretic*. Farrar, Straus and Giroux.

Safranksi. Rüdiger. 2009. *Romantik*. Fischer.

Safranksi. Rüdiger. 2017. *Goethe: Life as a Work of Art*. Liveright (Kindle Edition).

Santner. Eric L. 1990. *Friedrich Hölderlin: Hyperion and Selected Poems*. Continuum.

Saunders. Simon. 2000. *Clock watcher*. New York Times, March 26, 2000.

Saunders. Simon. et al. 2010. *Many Worlds: Everett, Quantum Theory and Reality*. Oxford University Press.

Schaffer. Jonathan. 2010. *Monism: The priority of the whole*. Philosophical Review 119 (1), pp. 31–76.

Schaffer. Jonathan. 2018. *Monism*. Stanford Encyclopedia of Philosophy (Winter). In: Zalta. Edward N. (ed.). *The Stanford Encyclopedia of Philosophy*.https://plato.stanford.edu/archives/win2018/entries/monism (accessed March 11, 2020).

Schefer. Christina. 2001. *Platons unsagbare Erfahrung*. Schwabe.

Schelling. Friedrich W. J. 2004. *First Outline of a System of the Philosophy of Nature*. SUNY Press.

Schelling. Friedrich W. J. 2013. *Ideen zu einer Philosophie der Natur*. Jazzybee Verlag (Kindle Edition).

Schilpp. Paul A. 1949. *Albert Einstein: Philosopher-Scientist*. Cambridge University Press.

Schlosshauer. Maximilian. 2008a. *Decoherence and the Quantum-to-Classical Transition*. Springer.

Schlosshauer. Maximilian. 2008b. *Lifting the fog from the north*. Nature 453 (39).

Schopenhauer. Arthur. 2010. *The World as Will and Representation* [Translated and edited by Judith Norman, Alistair Welchman & Christopher Janaway]. Cambridge University Press.

Schrödinger. Erwin. 1935a. *Discussion of probability relations between separated systems*. Mathematical Proceedings of the Cambridge Philosophical Society 31 (4), pp. 555–563.

Schrödinger. Erwin. 1935b. *Probability relations between separated systems*. Mathematical Proceedings of the Cambridge Philosophical Society 32 (3), pp. 446–452.

Schrödinger. Erwin. 1935c. *Die gegenwärtige Situation in der Quantenmechanik*. Die Naturwissenschaften 23, pp. 807–812.

Schrödinger. Erwin. 1952a. *Are there quantum jumps? Part I*. British Journal for the Philosophy of Science 3 (10) (August), pp. 109–123.

Schrödinger. Erwin. 1952b. *Are there quantum jumps? Part II*. British Journal for the Philosophy of Science 3 (11) (November), pp. 233–242.

Schrödinger. Erwin. 1992. *What Is Life?* Cambridge University Press.

Schrödinger. Erwin. 2014. *Nature and the Greeks and Science and Humanism*. Cambridge University Press.

Sears. Jane. 1952. *Ficino and the Platonism of the English Renaissance*. Comparative Literature 4 (3) (Summer), pp. 214–238.

Segre. Emilio. 2007. *From Falling Bodies to Radio Waves*. Dover.

Shelley. Mary. 1831. *Frankenstein, or the Modern Prometheus*. Henry Colburn & Richard Bentley.

Spinoza. Benedictus de. 1910. *Short Treatise on God, Man and His Well-Being*. A. & C. Black.

Spinoza. Benedictus de. 2017. *The Ethics*. Prabhat Prakashan (Kindle Edition).

Stoica. Ovidiu Cristinel. 2021. *Refutation of Hilbert space fundamentalism*. arXiv:2103.15104.

Strathern. Paul. 2007. *The Medici: Godfathers of the Renaissance*. Vintage (Kindle Edition).

Suh. H. Anna. 2013. *Leonardo's Notebooks*. Black Dog & Leventhal/Running Press (Kindle Edition).

Susskind. Leonard. 2006. *The Cosmic Landscape*. Little, Brown & Company.

Susskind. Leonard. 2009. *The Black Hole War*. Back Bay.

Susskind. Leonard. 2016. *Copenhagen vs Everett, teleportation, and ER=EPR*. arXiv:1604.02589.

Susskind. Leonard. 2017. *Dear Qubitzers, GR=QM*. arXiv:1708.03040.

Susskind. Leonard & Friedman. Art. 2015. *Quantum Mechanics: The Theoretical Minimum.* Basic.

Taylor. Thomas. 2016. *The Hymns of Orpheus: To Nature.* Bonificio Masonic Library.

Tegmark. Max. 1996. *Does the universe in fact contain almost no information?* Foundations of Physics Letters 9 (1), pp. 25–42. arXiv:quant-ph/9603008.

Tegmark. Max. 1997. *The interpretation of quantum mechanics: Many worlds or many words.* Fortschritte der Physik 46, pp. 855–862.

Tegmark. Max. 2000. *The importance of quantum decoherence in brain processes.* Physical Review E 61, pp. 4194–4206. arXiv:quant-ph/9907009.

Tegmark. Max. 2003a. *Parallel Universes.* In: Barrow. John D., Davies. Paul C. W., & Harper, Jr., Charles L. *Science and Ultimate Reality: From Quantum to Cosmos.* Cambridge University Press.

Tegmark. Max. 2003b. *Parallel Universes.* Scientific American 288 (5) (May), pp. 30–41.

Tegmark. Max. 2007. *Many lives in many worlds.* Nature 448, p. 23.

Tegmark. Max. 2008. *The mathematical universe.* Foundations of Physics 38, pp. 101–150. arXiv: 0704.0646 [gr-qc].

Tegmark. Max. 2009. *Many worlds in context.* In: Saunders et al. 2010, pp. 553–581, arXiv:0905.2182 [quant-ph].

Tegmark. Max. 2015a. *Consciousness as a state of matter.* Chaos, Solitons & Fractals 76, pp. 238–270. arXiv:1401.1219.

Tegmark. Max. 2015b. *Our Mathematical Universe.* Vintage.

Tegmark. Max & Wheeler. John Archibald. 2001. *100 years of the quantum.* Scientific American 284: 68–75 (February).

Terra Cunha. Marcelo O. Dunningham. Jacob A. & Vedral. Vlatko. 2006. *Entanglement in single particle systems.* arXiv:quant-ph/0606149.

Thorne. Kip. 2008. *John Archibald Wheeler 1911–2008.* Science 320, June 20, 2008, p. 1603. arXiv:1901.06623.

Tononi. Guido. 2008. *Consciousness as integrated information: A provisional manifesto.* Biological Bulletin 215, pp. 216–242.

Toole. Betty A. 1998. *Ada, the Enchantress of Numbers.* Pickering & Chatto.

Vaidman. Lev. 2016. *All is psi.* arXiv:1602.05025.

Van Helden. Albert. 1989. *Galileo Galilei: Sidereus Nuncius.* University of Chicago Press.

Van Raamsdonk. Mark. 2010. *Building up spacetime with quantum entanglement.* General Relativity and Gravitation 42, pp. 2323–2329.

Van Raamsdonk. Mark. 2016. *Lectures on gravity and entanglement.* arXiv: 1609.00026.

Varela. Francisco. 1995. *The Emergent Self*. In: Brockman. John. *The Third Culture*. Simon & Schuster, p. 209.

Verlinde. Erik. 2011. *On the origin of gravity and the laws of Newton*. Journal of High Energy Physics 29. arXiv: 1001.0785.

Wallace. David. 2012. *The Emergent Multiverse*. Oxford University Press (Kindle Edition).

Watts. Edward J. 2017. *Hypatia*. Oxford University Press.

Weizsäcker. Carl Friedrich v. 1971. *Die Einheit der Natur*. Carl Hanser.

Wheeler. John Archibald. 1990. *Information, Physics, Quantum: The Search for Links*. In: Zurek 1990, pp. 3–28.

Wheeler. John Archibald. 1996. *Time Today*. In: Halliwell, Perez-Mercader & Zurek 1994, pp. 1–29.

Wheeler. John Archibald. 2000. *"A practical tool," but puzzling too*. New York Times, December 12, p. F1.

Wheeler. John Archibald & Ford. Kenneth. 1998. *Geons, Black Holes and Quantum Foam*. W. W. Norton.

Wheeler. John Archibald & Zurek. Wojciech. 1983. *Quantum Theory and Measurement*. Princeton University Press.

Whitaker. Andrew. 2012. *The New Quantum Age*. Oxford University Press (Kindle Edition).

Whitehead. Alfred North. 1979. *Process and Reality*. Free Press.

Whiteley. Giles. 2018. *Schelling's Reception in Nineteenth-Century British Literature*. Macmillan.

Whitman. Walt. 2004. *The Complete Poems*. Penguin.

Wigner. Eugene P. 1995. *New Dimensions of Consciousness*. In: Mehra. Jagdish & Wightman. Arthur (eds.). 1995. *The Collected Works of E. P. Wigner, Volume VI: Philosophical Reflections and Syntheses*. Springer, pp. 268–273.

Wilczek. Frank. 2015a. *A Beautiful Question*. Penguin (Kindle Edition).

Wilczek. Frank. 2015b. *Physics in 100 years*, Physics Today 69 (4), p. 32.

Wilczek. Frank. 2016. *Physics in 100 years*. arXiv:1503.07735.

Witten. Edward. 2018. *Symmetry and emergence*. Nature Physics 14 (2), pp. 116–119. arXiv:1710.01791.

Wittmann. Marc. 2017. *Felt Time*. Massachusetts Institute of Technology Press.

Wittmann. Marc. 2018. *Altered States of Consciousness*. Massachusetts Institute of Technology Press.

Wolchover. Natalie. 2013a. *Is nature unnatural?* Quanta Magazine, May 24, 2013. https://www.quantamagazine.org/complications-in-physics-lend-support-to-multiverse-hypothesis-20130524 (accessed September 23. 2021).

Wolchover. Natalie. 2013b. *A jewel at the heart of quantum physics*. Quanta Magazine, September 17, 2013. https://www.quantamagazine.org/physicists

-discover-geometry-underlying-particle-physics-20130917 (accessed September 25, 2021).
Wolchover. Natalie. 2022. *A deepening crisis forces physicists to rethink structure of nature's laws*. Quanta Magazine. https://www.quantamagazine.org/crisis-in-particle-physics-forces-a-rethink-of-what-is-natural-20220301 (accessed April 3, 2022).
Wordsworth. William. 2001. *The Prelude of 1805*. Global Language Resources.
Wulf. Andrea. 2015. *The Invention of Nature*. John Murray.
York. Michael. 2003. *Pagan Theology*. New York University Press.
Zeh. H. Dieter. 1967. *Probleme der Quantentheorie*. Unpublished. http://www.rzuser.uni-heidelberg.de/~as3/ProblemeQT.pdf (accessed September 18, 2021).
Zeh. H. Dieter. 1970. On the interpretation of measurement in quantum theory. Foundations of Physics 1 (1), pp. 6976.
Zeh. H. Dieter. 1993. There are no quantum jumps, nor are there particles! Physics Letters A 172, pp. 189–195.
Zeh. H. Dieter. 1994. *Warum Quantenkosmologie?*. Unpublished talk. http://www.rzuser.uni-heidelberg.de/~as3/WarumQK.pdf (accessed September 18, 2021).
Zeh. H. Dieter. 2004. The wave function: It or bit? In: Barrow. J. D. Davies. P. C. W. & Harper. C. L. Jr. (eds.). *Science and Ultimate Reality*. Cambridge University Press, pp. 103–120. arXiv:quant-ph/0204088.
Zeh. H. Dieter. 2006. Roots and fruits of decoherence. In: Duplantier. B. Raimond. J.-M. & Rivasseau. V. (eds.). *Quantum Decoherence*. Birkhäuser, pp. 151–175. arXiv:quant-ph/0512078.
Zeh. H. Dieter. 2007. *The Physical Basis of the Direction of Time*. Springer.
Zeh. H. Dieter. 2012a. *Physik ohne Realität: Tiefsinn oder Wahnsinn?* Springer (Kindle Edition).
Zeh. H. Dieter. 2012b. Open questions regarding the arrow of time. In: Mersini-Houghton. Laura & Vaas. Rüdiger (eds.). *The Arrows of Time*. Springer, pp. 205–217.
Zeh. H. Dieter. 2018. The strange (hi)story of particles and waves. arxiv: 1304.1003v23.
Zurek. Wojciech. 1990. *Complexity, Entropy and the Physics of Information*. Westview.
Zurek. Wojciech. 1991. Decoherence and the transition from quantum to classical. Physics Today 44, pp. 36–44.
Zurek. Wojciech. 2009. Quantum Darwinism. Nature Physics 5, pp. 181–181.
Zurek. Wojciech. 2014. Quantum Darwinism, classical reality and the randomness of quantum jumps. Physics Today, October 2014, p. 44.

INDEX

accelerators, 5–6, 192–193, 202, 283
Adams, Ansel, 181
AdS/CFT, 235, 242–243, 247
Aeschylus, 112
Age of Entanglement, The (book), 40
Aguirre, Anthony, 213, 238
Aikenhead, Thomas, 169
Albert, David, 32, 260
Albert, Karl, 52
Alcibiades, 129
Alcuin of York, 125
Alexander the Great, 118
al-Kindi, Abu Yusuf, 121
"All Is Psi," 214
Almagest (book), 154
Almheiri, Ahmed, 236
Altered States of Consciousness (book), 268
Amalric of Bena, 133, 141
amplituhedron, 244–245
animism, 49
ant perspective, 220–224, *221*
"anthropic argument," 195–196, 262
Anti-de Sitter (AdS), 235, 242–243, 247
Aquinas, Thomas, 133–134, 149
Archimedes, 118
Archytas, 112
Argyropoulos, John, 136
Aristarchus of Samos, 118
Aristophanes, 112, 129, 149
Aristotle, 112–113, 120, 131–133, 157
Arkani-Hamed, Nima, 197, 216–218, 244–245

Arndt, Markus, 90
Aspect, Alan, 96
Assayer (book), 157
Assmann, Jan, 50, 104–106, 173
Astrophil and Stella (book), 166
atomic orbits, 18–19, 83, 204
Augustine of Hippo, 116–117, 125–126

Babbage, Charles, 183
"baby universes," 2, 195–197, 212, 239–240
Back to the Future (film), 29
Bacon, Francis, 149
Ball, Philip, 259–261
Barbarossa, Frederick, 130–131
Barbour, Julian, 226–229
Barnes, Jonathan, 107
Barrett, Jeffrey, A., 260
Beautiful Mind, A (film), 68
Beautiful Question, A (book), 188
beauty, 127, 147–190, 194–197, 209–212, 218
Beethoven, Ludwig van, 173, 279
Bekenstein, Jacob, 233, 240
Bell, John Stewart, 47, 62, 92–95, 104, 123
Bellarmine, Cardinal, 143
Benivieni, Girolamo, 138, 166
Berenstain, Nora, 53
Berlin University, 180
Berliner Tageblatt (newspaper), 57
Bertlmann, Reinhold, 93

Bessarion, 136–137, 154
Bible, 115–116, 122, 129
Big Bang, 2, 13, 278
Big Picture, The (book), 200
bird perspective, 97, 97–102, 128, 212–213, 222, 269–271
Birth of Venus, The (painting), 3, 138
"bit," 37, 241, 276–277, 281
black holes
  entanglement and, 217–218, 240–243
  general relativity and, 232–246
  mysteries of, 232–237
  naming, 11
  properties of, 232–237, 240–243, 258
  string theory and, 235–241
  thermodynamics and, 233, 240–241, 263, 275
  Unruh effect and, 88, 233–234
  wormholes and, 11, 237–240, 242
Blackmore, Susan, 268
Boethius, Anicius Manlius, 120, 156
Bohm, David
  entanglement and, 44–45, 47–48, 213
  hidden variables and, 68–71
  many worlds and, 80
  quantum mechanics and, 44, 47–48, 68–71, 92, 95
Bohr, Harald, 17
Bohr, Niels
  complementarity and, 27–29, 40, 52–55, 76–77, 114, 133, 137, 178, 234–235, 250
  entanglement and, 42
  illustration of, 9
  multiverse and, 73–77
  physics and, 10, 178–179
  projector reality and, 15–16, 133
  quantum mechanics and, 46–47, 53–58, 61, 69, 92, 102, 248
  quantum physics and, 22–38, 256
  wave function and, 43–44, 70, 226, 281–282
Bondi, Hermann, 165
"Book of Nature," 150, 156–160
Borgia, Cesare, 139
Born, Max, 20, 22–23, 30
Botticelli, Sandro, 3, 8, 138, 153
Botvinick, Matthew, 264
Bousso, Raphael, 213, 238, 252
Boyle, Robert, 163

Bracciolini, Poggio, 135
Bragg, William, 30
Brahe, Tycho, 158–159
"Brahma," 51–52
Brillouin, Léon, 30
Bringing Up Baby (film), 14, 29
British Journal for the Philosophy of Science (journal), 86
Brown University, 222
Brunelleschi, Filippo, 135
Bruno, Giordano, 103, 104, 122, 140–145, 151–152, 155–160, 166–169, 186
"Bucky Balls," 90
Burgess, Cliff, 216
Burned Alive (book), 142
Burnet, John, 107
Byrne, Peter, 75–76, 231
Byron, Lord, 183

California Institute of Technology, 10, 40, 42, 83, 200
Capra, Fritjof, 54
Carabine, Deirdre, 123–125, 127–128
Carl Eugen, Duke of Württemberg, 175
Carroll, Sean, 40, 200, 214, 248, 255
Carter, Howard, 50
Catholic University of America, 67
"causal diamond," 252
causality, 22, 26, 31, 43–46, 57–61, 144
Cause, Principle and Unity (book), 140
cave allegory, 116, 179, 220, 249, 278, 282
celestial music, 156–160, 188
celestial objects, 1–3, 168
CERN, 92, 192–196, 288
Chalmers, David, 256
Champion, Justin, 169–170
Charlemagne, 125
Charles the Bald, 125–127
chicken-egg problem, 265–266, 266, 280, 280–281
Clairmont, Claire, 183
classical mechanics, 165, 168–169, 175, 181
classical physics, 22, 29, 40, 68–70, 212, 227, 258, 290
classical reality, 37, 40, 80–81, 98–101, 250–257, 264
Clauser, John, 92, 94, 96
"Clauser-Horne-Shimony-Holt (CHSH) inequality," 94, 96

Index

clockwork universe, 165–169, 175–176, 181
Cohen, Jonathan, 264
Cohen, Leonard, 278
Cold War, 67–68, 76
Coleridge, Samuel Taylor, 4, 180, 184, 188
Columbus, Christopher, 137–138
complementarity
    Bohr and, 27–29, 40, 52–55, 76–77, 114, 133, 137, 178, 234–235, 250
    explanation of, 27–29
    projector reality and, 52–55
    quantum cosmology and, 40, 250
    quantum reality and, 52–53
    quantum weirdness and, 40–41
Compton, Arthur, 30
"Computing Space," 276
Comte, Auguste, 203–204
Conference on the Role of Gravitation in Physics, 74, 224
conformal field theory (CFT), 235, 242–243, 247
Conscious (book), 260
"conscious One," 249, 249–271, 266
consciousness, altered state of, 267–271, 284
"Consciousness as a State of Matter," 255
conservation, 57, 162, 181–182, 210, 231
Consolation of Philosophy, The (book), 120
Constantine the Great, 120
"Constructor Theory," 245–246
Cook, James, 4, 172
Copenhagen interpretation, 15–17, 23–34, 42–66, 73–104, 123, 209, 238
Copernicus, Nicolaus, 75, 140, 142, 144, 147, 153–156, 159
cosmic inflation, 2, 195, 212–215, 238–239
cosmos
    "all is One" and, 105, 114, 121
    beauty and, 150–189
    "One to the rescue" and, 200–211
    particle physics and, 283–289
    projector reality and, *14*, 14–15, *15*
cosmotheism, 104
Coustos, John, 170–171
Crick, Francis, 199
Critique of Pure Reason (book), 175–176
Csikszentmihalyi, Mihaly, 267
Cudworth, Ralph, 166–167, 171

Curie, Marie, 30
Cusanus, 135–137, 140, 145. *See also* Nicholas of Cusa
Cyril of Alexandria, 119

da Vinci, Leonardo, 138, 147, 151–153, 155, 188
dark energy, 2, 193–195, 215–218, 283
dark matter, 2, 160, 193–194
Darwin, Charles, 185
Darwin, Erasmus, 185, 187
Darwinism, 187, 253–254, 286
David of Dinant, 141
Davy, Humphry, 182, 184
d'Azyr, Félix Vicq, 174
de Broglie, Louis, 18, 21, 30, 45, 66–71, 95–96
De musica (book), 156
Declaration of Independence, 4, 173
decoherence
    bird perspective and, 97, 97–102, 128, 212–213, 222, 269–271
    "causal diamond" and, 252
    consequences of, 98, 98–99
    discovery of, 84, 87–92, 96–97, 227
    entanglement and, 65–66, 100–103, 136–137, 186
    frog perspective and, 97–102, 99, 180, 212–213, 222, 231, 253, 259, 269–271
    pioneers of, 77, 84, 92, 250
    "preferred basis problem," 251–255, 262–265
    process of, 7, 84–98, 99, 99–103, 195, 213, 222, 227–231, 242, 258–260, 264, 269–271, 278
    role of, 87–99, 99
    superpositions and, 87–91, 258–260
    theory of, 65, 77, 125–129, 180, 250–253
Decoherence and the Quantum-to-Classical Transition (book), 90
deism, 150, 169
Delft University, 90
Desaguliers, John Theophilus, 170
Descartes, René, 149
d'Espagnat, Bernard, 52, 92–95, 163
determinism, 21–22, 43, 58–61, 98, 133, 165, 259, 264

Deutsch, David, 78, 100–101, 245–246, 275
DeWitt, Bryce
  many worlds and, 71, 75, 81
  parallel universes and, 71–72
  quantum gravity and, 85
  quantum mechanics and, 71–75, 81, 85, 95–96, 224–228
  Wheeler-DeWitt equation, 11, 224–228
DeWitt-Morette, Cécile, 73–75
Diamond, Jared, 49
Dionysius the Areopagite, 116, 122–124, 126, 140
Dirac, Paul, 30, 226–227
Discourses and Mathematical Demonstrations Relating to Two New Sciences (book), 160–161
Divine Names, The (book), 123
Division of Nature, The (book), 127
DNA, 107, 199
dualism, 116–117, *117*, 126–132, 139, 287–288
"Duhem-Quine thesis," 25

Early Greek Philosophy (book), 107
Eckhart, Meister, 122, 130–135, 140, 145
ego, 176–178, 264–266, 284
Ego Tunnel, The (book), 264
Ehrenfest, Paul, 31
Eighty Years' War, 161
Einstein, Albert
  entanglement and, 6, 38–48, 236–241
  EPR paradox, 42–47, 66, 71, 78, 91–94, 214, 237–241, 261
  general relativity and, 11, 17–19, 29–31, 44–45, 66–73, 184, 203, 223–226, 231–232, 247, 285
  gravity and, 220, 223
  illustration of, 39
  multiverse and, 73
  quantum mechanics and, 11, 42–47, 55–69, 94, 163–164, 220–226, 247–248
  quantum physics and, 17, 23–25, 42
  science of, 8
  seminar by, 66, 69
  space-time and, 223–224, 231–232, 285
  special relativity and, 29, 44–45, 58–59, 220, 223
  universe and, 194
"Einstein Attacks Quantum Theory," 43
Einstein-Podolsky-Rosen (EPR) paradox, 42–47, 66, 71, 78, 91–94, 214, 237–241, 261
electricity, 6, 178, 181–184, 210
electrochemistry, 181–184
electrodynamics, 11, 184, 188, 210, 224
electromagnetism, 16, 30, 86, 181–184, 210
Elements (book), 118–119
Elements of Pantheism (book), 286
Elizabeth II, Queen, 276
elliptic orbits, 155, 159–160
Ellis, George, 196
emergence, 12, 102, 125–128, 176, 195–203, 228–243, 263–265
Emerson, Ralph Waldo, 181
Emperor's New Mind, The (book), 258
empiricism, 150, 179, 184
End of Time, The (book), 227
Engels, Friedrich, 76
Enneads, The (book), 112, 121, 129
entanglement
  black holes and, 217–218, 240–243
  consequences of, 43–47, 91, 213–217, 219–248, 270
  decoherence and, 65–66, 100–103, 136–137, 186
  explanation of, 6–7, 38–52, 71–96, 128–129, 168, 179–180, 201–220, 236–260
  as glue of worlds, 40–48
  phenomenon of, 40–48, 209, 243–244
  quantum cosmology and, 128–129, 213
  quantum weirdness and, 7, 40–41
  role of, 91–96
  space-time and, 219–248
  superpositions and, 72, 215, 242–243, 270
entropy, 205–207, *206*, 229–233, 241–242
Epistemological Letters (journal), 96
ER=EPR principle, 237–240, 261
Eriugena, John Scotus, 122, 125–130, 133, 136, 140, 163, 174, 284
Erman, Paul, 186–187
"Essay on the Geography of Plants," 174

Ethics (book), 163–165
Euclid, 118–119
Euripides, 112
European Organization for Nuclear Research (CERN), 92, 192–196, 288
European Space Agency, 5, 15
Everett, Hugh
 EPR paradox and, 71
 illustration of, 65
 many worlds and, 2–3, 61–67, 71–89, 96–101, 195, 212–213, 224, 238–239, 257–261
 parallel universes and, 2–3, 71–72, 99–100, 238–239
 physics and, 11
 projector reality and, 16
 quantum mechanics and, 63, 66–70, 92–95, 99–101, 104, 143, 224, 226, 246–248, 270
 time and, 231
 "uglyverse" and, 212
 wave function and, 224, 226, 246–247
Everett, Mark, 231

Everett interpretation, 71–77, 78, 78–80, 96–101. *See also* many worlds
Expulsion of the Triumphant Beast, The (book), 141

Faerie Queene, The (book), 166
Falke, Bernd, 84
Fall of Man, 125–129, 284
Faraday, Michael, 8, 147, 182, 184
Ferguson, Kitty, 160
Fermi, Enrico, 95
Feynman, Richard, 10–11, 13, 56, 74, 224, 227, 245
Fichte, Johann Gottlieb, 176–178
Ficino, Marsilio, 138, 147–156, 166–167, 171, 174, 188
film roll
 complementarity and, 52–55
 on-screen reality and, 19, 29, 60–61, 97, 113, 250, 271, 275–279, 280, 280–283, 282
 perspective and, *14*, 14–16
 projector reality and, *14*, 14–16, *15*, 19, 28–39, 61–63, 88, 275

fine-tuning, 214, 218, 283. *See also* naturalness
firewall paradox, 236–240
First Outline of a System of the Philosophy of Nature (book), 179
Flasch, Kurt, 125
"flow," 267
Ford, Ken, 38
Forster, Georg, 172–173, 177–178, 199
Foundational Questions Institute (FQXi), 269
Foundations of Physics (journal), 95
"Foundations of Quantum Mechanics in the Philosophy of Nature," 60
Fowre Hymnes (book), 166
Francis I, King, 153
Frankenstein, *147*, 180–183
Frederick II, Emperor, 130–131
Frederick IX, King, 54
Freedman, Michael, 246
Freire, Olival Jr., 94–95
French Revolution, 171–176
frog perspective, 97–102, 99, 108, 180, 212–213, 222, 231, 253, 259, 269–271, 286
fundamentality, 145, 194–208, *208*, 209–232, 240–244

galaxies, 1–3, 72, 232, 278
Galen, 118
Galilei, Galileo, 140, 142–144, 147–151, 154–158, 160–161
Galilei, Vincenzo, 156–157
Gedanken experiments, 30
General Relativity and Gravitation (journal), 242
general relativity theory
 discovery of, 285
 Einstein and, 11, 17–19, 29–31, 44–45, 66–73, 184, 203, 223–226, 231–232, 247, 285
 quantum mechanics and, 11, 42–47, 55–62, 94, 223–226, 247–248
 quantum physics and, 29–31, 188
 special relativity and, 29, 44–45, 58–59, 220, 223
George I, King, 169
Gilder, Louisa, 40
Giudice, Gian, 192, 194
Gleick, James, 165

God
  concept of, 4, 8
  monotheism and, 104–107, *106*, 169–172
  nature and, 160–174, 186
  "One" and, 4, 8, 43, 46, 50, 54, 67–69, 104–107, 114–145, 154–156, 169–172
  pantheism and, 49–50, 104–107, 115, 134, 141, 149–150, 169–172, 180–181
  perspectives, 180–181, *221*, 221–223, 286
  "unknown God," 122–124
God perspective, 180–181, *221*, 221–223, 286
Goddu, André, 154
Gödel, Escher, Bach (book), 265
"God's eye," *221*, 221–223
Goebbels, Joseph, 187
Goethe, Johann Wolfgang von, 4, 8, 147, 171–174, 177, 180–182, 186–188, 287–288
golden ratio, 152–154, 160
Gomes, Henrique, 214
"good and evil," 116–117, *117*, 126–129. *See also* dualism
Google, 11
Gottschalk, 125–126, 130
"GR=QM," 248
Graham, Neill, 75
Grant, Cary, 14
Gravity Research Foundation, 242
Greek philosophy, 107, 111–116, 123, 134–136, 148, 283–284
Gregory IX, Pope, 131–132
Gregory of Nyssa, 116
Griffiths-Dickson, Gwen, 49
Gröblacher, Simon, 90
Grove, William, 182
Gutenberg, Johannes, 148

Haeckel, Ernst, 185–187
hallucinogenics, 269–270
Hamsun, Knut, 187
Harari, Yuval Noah, 197–198, 200
Harmonies of the World (book), 159
Harris, Annaka, 260, 264
Harrison, Paul, 286

Hawking, Stephen, 4, 11–12, 78, 196, 233–242, 258, 263, 276
Hegel, Georg Wilhelm Friedrich, 128, 174, 177–180, 186–187
Heine, Heinrich, 180, 188
Heisenberg, Werner
  "Heisenberg cut" and, 257–258
  Helgoland and, 19, 225
  illustration of, 9
  projector reality and, 15–16
  Pythagorean ideas and, 112
  quantum mechanics and, 18–32, 42–45, 53–62, 86, 102, 198, 277
  religion and science, 55, 133
  seminar by, 57–58
  symmetry and, 210
  uncertainty principle and, 25–28, *27*, 43–45, 50
  wave function and, 36, 43–44
"Heisenberg cut," 257–258
"Hen Kai Pan," 167–168, 174–175
Henry III, King, 142
Hepburn, Katharine, 14
Heraclitus, 3, 6–7, 76, 111, 140, 178, 221–222
Herder, Johann Gottfried, 171
Hermann, Grete, *39*, 57–62, 210
Herodotus, 112
Hesiod, 108
hidden reality, 9, 13, 23, 39, 136, 281–282, *282*
hidden variable theories, 45–46, 58–62, 68–71, 93
hierarchies, 151, 203–204, *208*, 208–209
hierarchy problem. *See* fine-tuning; naturalness
Higgs mass, 189, 192–195, 202, 215–218, 283
Hilbert, David, 35, 58
Hilbert space, *35*, 35–37, 179, 210, 255
Hippocrates, 112
Hitler, Adolf, 10, 42, 59, 82, 198
Hofstadter, Douglas R., 265
Hölderlin, Friedrich, 174–177, 180, 186
holism, 214, 269, 284
Holmes, Richard, 182, 187
holographic principle, 235–236, 241–242, 247
Holy Trinity, 115–116, 142

Hooft, Gerard 't, 234
Hooke, Robert, 165
Hossenfelder, Sabine, 196
Hu, Wayne, 160
Humboldt, Alexander von, 3, 153, 172–174, 181, 185
Hund, Friedrich, 57–58
Huxley, Aldous, 52, 268
Huygens, Christiaan, 161–163, 165
hydrogen bomb, 10, 38
Hylleraas, Egil A., 46
Hypatia of Alexandria, 118–119

I Am a Strange Loop (book), 265
IBM, 11
Ibn Sina, Abu, 121
inflation of universe, 2, 195, 212–215, 238–239
information,
    biology and, 199
    black holes and, 232–236, 240-241
    brain and, 268
    complementarity and, 28
    decoherence and, 85, 87–88, 101–102, 207, 231, 250, 253, 269
    entanglement and, 5, 45, 92, 94, 238
    entropy and, 205, 207, 229, 241
    hidden variables and, 58
    integrated, 261–263
    language and, 197–198
    matter and, 37–38, 199, 277–279, 280, 285 280–283, 282, 289
    quantum, 11, 40, 93, 215, 244, 248, 255, 275, 283
    quantum measurement and, 251–254
    quantum reality and, 14
    spacetime and, 241, 243
    theory, 204–205, 276
Innocent III, Pope, 132
Institute for Advanced Study, 42, 62, 68–69, 163, 197, 224, 247
Institute for Political Science, 84
integrated information, 261–263
Intel, 11
International School of Physics, 95
"Interpretation," 260
Ion of Chios, 110
Isaacson, Walter, 152

Isis, 51, 102, 111, 116, 135, 138, 141, 169, 173–174, 274, 282. *See also* Neith
Israel, Jonathan, 163
"It from Bit," 37, 276, 281
"It from Qubit," 244

Jacob, Margaret, 169
Jacobi, Friedrich Heinrich, 171–172, 188
Jacobsen, Anja Skaar, 77
Jacobson, Ted, 240–241
Jaksland, Rasmus, 244, 247
Jammer, Max, 66, 75, 164
Jaspers, Karl, 128
Jensen, Hans, 83, 95
Jesus, 115–116, 131, 142, 170, 203
Joos, Erich, 86, 213, 250
Jordan, Pascual, 20–21, 55
Joshua (boat), 266–267

Kahn, Charles H., 116
Kaiser Wilhelm Institute for Physics, 17
Kaiser Wilhelm Society, 42
Kant, Immanuel, 57, 175–176, 178–179
Kennedy, Katherine, 270
Kepler, Johannes, 112, 140–142, 151, 155–160, 165, 175, 188
Kerouac, Jack, 181
Keynes, John Maynard, 165–166
Kiefer, Claus, 40, 227–231
Kierkegaard, Søren, 128
Koch, Christof, 262–263
Konrad of Marburg, 132
Korean War, 74
"Kosmos," 181
Kristeller, Paul Oskar, 138, 149, 151, 156
Kuhn, Thomas S., 182

landscape (string theory), 195, 207–208, 211, 238
Lao-tzu, 51, 54, 110, 114
Large Hadron Collider (LHC), *191*, 192–193
Lau, Darrell, 51
Leibniz, Gottfried W., 163, 247
Leo X, Pope, 138–139
leptons, 194, 210–211
Lessing, Gotthold Ephraim, 171–172
Letters to Serena (book), 169
Lévy-Leblond, Jean-Marc, 79–80

Liebig, Justus von, 187
living organism, 90, 114, 186, 198–200
Livio, Mario, 158, 160
Lloyd, Seth, 275–279
locality, 46, 93–96, 238–239, 245, 253–254
Lockwood, Michael, 81, 89, 255, 260, 262–263, 270
Loewer, Barry, 260
loop quantum gravity, 214, 228
Lost in Math (book), 196
Louis the Pious, 126
Lovelace, Ada, 183
LSD, 266–270
Lucas, Gary, 67
Lucretius, 135, 137, 169
Lumma, Dirk, 61
Luther, Martin, 139

macrostate, 205–207, 230
"Mad-Dog Everettianism," 214
Magic Flute, The (opera), 3, 173
magnetism, 6, 89, 178, 181–184, 210
Magnus, Albertus, 131, 133
Mahadevan, Telliyavaram, 51
Maldacena, Juan, 235, 238–240, 242
Maldacena's conjecture, 235, 238, 242
Mandelbrot set, 278
Manichaeism, 116–117, 132
many worlds
    Bohm and, 80
    DeWitt and, 71, 75, 81
    Everett and, 2–3, 61–67, 71–77, 78, 78–89, 96–101, 195, 212–213, 224, 238–239, 257–261
    Pythagorean ideas and, 142
    Wheeler and, 72–76, 80–81
Many-Worlds Interpretation of Quantum Mechanics, The (book), 75
Marletto, Chiara, 245–246
Marolf, Donald, 236
Martell, Karl, 121
Martínez, Alberto, 140, 142, 155
Marx, Karl, 76, 170, 187
Massachusetts Institute of Technology (MIT), 96, 222, 243, 275
Massimi, Michela, 179, 186
Mathematical Foundations of Quantum Mechanics (book), 58, 69
matrix mechanics, 20, 20–22, 26–28

matter
    birth of, 85–88
    consciousness and, 265–266, 266
    dark matter, 2, 160, 193–194
    defining, 7–8
    explanation of, 7–8, 85–88
    information and, 37–38, 199, 277–279, 280, 280–283, 282, 289
Maxwell, James Clerk, 73, 147, 184, 210
Mayer, Maria Goeppert, 83, 95
McGinnes, Jon, 121
McGuire, James, 166–168
Medici, Cosimo de', 137–138
Medici, Giovanni de', 138–139
Medici, Lorenzo de', 138–139
Medici, Piero de', 139
Mermin, David, 56
Metzinger, Thomas, 264–265, 268
Michelangelo, 138
Microsoft, 11
microstate, 205–207, 230
Milky Way, 1–2, 232
Mind, Brain and the Quantum (book), 255
Minkowski, Hermann, 220, 223
Mirandola, Pico della, 138–139, 167, 171
Misner, Charles, 69–70, 73–77, 233
"Mission of Moses, The," 173
Moitessier, Bernard, 266–267
momentum, 26–29, 27, 43–45, 149, 161–162, 210, 214, 225
Monet, Claude, 181
monism
    "all is One" and, 3, 8, 39, 48–56, 102, 104–108, 105–106
    beauty and, 146–174, 177–183, 186–189
    "conscious One" and, 256, 263–264
    diplomat of, 135–141
    explanation of, 3
    history of, 103–145
    "One is all" and, 67–68, 79–81, 111–116
    "One to the rescue" and, 191, 201, 207–214
    "space and time" and, 219–220, 231, 244–248
    "struggle for One" and, 103–145
    "unknown One" and, 282–289
monotheism, 104–107, 106, 116, 141, 169–172, 288

Morgenstern, Oskar, 68
Moses, 105, 116, 167, 170, 173
Mozart, Wolfgang Amadeus, 3, 8, 173
Muir, John, 181
multiverse
    discovery of, 71–75
    explanation of, 2–3, 71–85
    illustration of, 78
    string theory and, 238–239
    "uglyverse" and, 194–197, 212
    universe and, 2–3, 71–77, 78, 78–85, 194–197, 212–213

NASA, 5, 11, 15
Nash, John Forbes, 68
National Security Agency, 68
naturalness, 203–204, 209–212, 216–218
Nature (journal), 77
"Nature's God," 4
Neith, 50–51
Nelson, Leonard, 57, 60, 62
Neoplatonism, 112–113, 116–117, 121–122, 127, 151, 154. *See also* Platonism
New Scientist (magazine), 4, 196
New York Times (newspaper), 43, 275
Newton, Isaac
    clockwork universe and, 165–169, 175–176, 181
    Hugh Everett and, 73
    physics and, 8, 22, 29, 262–263
    scientific theories and, 147
    work by, 159–160, 165–171, 175–176, 181, 188–189, 210
Newtonian mechanics, 159–160, 165–171, 181
Newtonian physics, 8, 22, 29, 262–263
Nicholas of Cusa, 122, 135–141, 145, 151, 154, 158
Nixey, Catherine, 117–118
Noether, Emmy, 57, 59, 62, 210
Nomura, Yasunori, 213, 238
Novara, Domenico Maria da, 153–154
nuclear bombs, 10
nuclear models, 83–84
nuclear wars, 68, 82
nuclear weapons, 10, 58, 62, 68

observer, conscious, 12, 84, 249, 249–271, 266
"Ode to Joy," 173
Odoacer, 120
Oldenburg, Henry, 164–165
"On First Philosophy," 121
"On Granite," 173
On Learned Ignorance (book), 136
"On Nature," 111, 114
On the Nature of Things (book), 135
On the Revolutions of the Heavenly Spheres (book), 153–155
"One"
    "all is One," 4, 6–9, 39–63, 65–66, 87–102, 104–108, 113–114, 125–133, 145, 221, 269, 282–285
    beauty and, 127, 147–190, 194–197, 209–212, 218
    "conscious One," 249–271
    "hidden One," 9–38, 50, 65
    "One is all," 65–102, 111–116
    science and, 127–145, 147–190, 218–219, 275–276
    struggle for, 103–145
    "to the rescue," 191–218
    understanding, 1–8, 274–290
    "unknown One," 273–290
"One and All," 173–174
"One is all," 65–102, 111–116
on-screen reality
    Copenhagen interpretation and, 104
    film roll and, 19, 29, 60–61, 97, 113, 250, 271, 275–279, 280, 280–283, 282
    projector reality and, 19, 29, 37, 133, 163, 250, 271, 275–281
"Open Questions Regarding the Arrow of Time," 230
Oppenheimer, Robert, 68
Opticks (book), 167–168
Oration on the Dignity of Man (book), 138
orbits
    atomic orbits, 18–19, 83, 204
    elliptic orbits, 155, 159–160
    planetary orbits, 75, 109, 128, 158–160, 165, 168, 175, 223
Orestes, 119
Ornes, Stephen, 89
Orpheus, 108, 156, 167
Orphism, 108, 110–111
Ørsted, Hans Christian, 182–184
Ouellette, Jennifer, 244

Pacioli, Luca, 152–153
Page, Don, 263
Pan (book), 187
pantheism, 49–50, 104–107, 115, 134, 141–150, 169–172, 180–187, 286
Pantheisticon (book), 170
parallel universes, 2–3, 71–72, 99–100, 238–239
Parmenides, 111, 114, 137, 140, 221–222
Parmenides (book), 112–114, 154, 199
Part and the Whole, The (book), 54
particle physics
    explanation of, 4–11
    in trouble, 192–194
    understanding, 4–12, 107, 202–220, 244–248, 283–289
particles
    properties of, 16, 21–24, 33, 35–37
    spin of, 94, 193, 216, 270, 276–277
    universe and, 4–5, 16, 21–24, 29, 33, 35–37, 282–289
    waves versus, 16, 21–24, 29, 32, 55
Paul III, Pope, 155
Paul the Apostle, 115–116
Pauli, Wolfgang, 19–20, 22, 26–28, 30, 55, 93
Penrose, Roger, 258
Perimeter Institute, 217
Periphyseon (book), 129–130
perspective
    ant perspective, 220–224, *221*
    bird perspective, 97, 97–102, 128, 212–213, 222, 269–271
    film roll and, *14*, 14–16
    frog perspective, 97–102, 99, 180, 212–213, 222, 231, 253, 259, 269–271, 286
    God perspective, 180–181, *221*, 221–223, 286
    projector reality and, *14*, 14–16
    quantum reality and, *13*, 13–16
Petersen, Aage, 56, 69, 73–74
Phidias, 112
Philip II, King, 161
Philo of Alexandria, 116, 123–124
Philolaus, 110, 112
Philolaus (book), 112
Philosophiae naturalis principia mathematica (book), 165

Philosophical Transactions of the Royal Society (journal), 164
"Physical Science and the Study of Religions," 54
"Physics from Scratch," 214
Physics Today (journal), 95–96
Planck, Max, 30, 42, 66, 274
Planck scale, 12, 228, 239, 274
planetary orbits, 75, 109, 128, 158–160, 165, 168, 175, 223
planetary system, 1–3, 17, 72, 140, 154, 160, 232, 278
plasma, 2, 6, 160
Plato, 3, 32, 34, 36–38, 39, 76, 102, 111–122, 125–133, 136–142, 147–158, 166–167, 179, 185, 199, 211, 220–222, 227, 249, 277–278, 282
"Platonia," 227
Platonic Paradigm, 102
Platonism, 111–122, 126, 130, 142, 148–154, 166–167. *See also* Neoplatonism
Pletho, Gemistus, 136–138, 154
Plotinus, 112–113, 119, 121, 129, 138, 174
Plutarch, 50, 154
Podolsky, Boris
    entanglement and, 41–47, 71, 78
    EPR paradox, 42–47, 66, 71, 78, 91–94, 214, 237–241, 261
Polchinski, Joseph, 236
Poliziano, Angelo, 139
Polo, Marco, 131
polytheism, 49, 104–105, 108
"preferred basis problem", 251–255, 262–265
Princeton University, 63, 67–69
Principia (book), 166, 168
"Problems in Quantum Theory," 261
Programming the Universe (book), 275
projector reality
    complementarity and, 52–55
    entanglement and, 247–248
    film roll and, *14*, 14–16, *15*, 19, 28–39, 61–63, 88, 275
    on-screen reality and, 19, 29, 37, 133, 163, 250, 271, 275–281
    perspective and, *14*, 14–16

quantum reality and, 13–17, *15*, 32, 39–40, 52–53, 123–124, 163, 250, 271
Prometheus (book), 172
Prussian Academy of Science, 17
psychoactive drugs, 267–270
Ptolemy, 153–154
Pyramid Texts, 50–51
Pythagoras, 110–112, 127, 155–156, 167
Pythagorean ideas, 110–112, 142–143, 148–159, 171
Pythagoreanism, 142–143

"Q-Bit," 276–277
quanta, 16–20, 30, 86, 89, 202, 208–209, 236
quantum computing, 78–79, 274–277
quantum cosmology
   celestial music and, 156–160, 188
   complementarity and, 40, 250
   entanglement and, 128–129, 213
   explanation of, 1–11
   "One is all" and, 70–102
   on-screen reality and, 282–283
   Pythagorean ideas and, 142–143
   space-time and, 219–231
quantum Darwinism, 253–254
quantum entanglement. *See* entanglement
quantum factorization, 79, 98, 98–99, 125, 180, 251–265
quantum field theory, 184, 211–218, 234–235, 242–245
quantum foam, 12, 225–226
quantum gravity, 69–70, 85, 210–248, 255–260, 275–278, 289
quantum holism, 214, 269, 284
quantum information, 11, 40, 93, 215, 244–248, 255, 275, 283–286
quantum jumping, 18, 21–24, 86
quantum mechanical wave, 86, 97, 136, 224–225, 257, 281
quantum mechanics
   Bohm and, 44, 47–48, 68–71, 92, 95
   Bohr and, 46–47, 53–58, 61, 69, 92, 102, 248
   description of, 2–11
   DeWitt and, 71–75, 81, 85, 95–96, 224–228
   Einstein and, 11, 42–47, 55–69, 94, 163–164, 220–226, 247–248

   entanglement and, 6–7, 38–52, 65–66, 71–96, 100–103, 128–129, 136, 168, 179–180, 186, 201–220, 236–260, 270
   Everett and, 63, 66–70, 92–95, 99–101, 104, 143, 224, 226, 246–248, 270
   Heisenberg and, 18–32, 42–45, 53–62, 86, 102, 198, 277
   hidden variables and, 45–46, 58–62, 68–71, 93
   matrix mechanics, 20, 20–22, 26–28
   understanding, 2–11, 16–17
   universe and, 2–11
   von Neumann and, 34–35, 58–63, 66–71, 95, 256–257
   Wheeler and, 10–13, 28, 37–38, 52, 63, 66, 77, 82–83, 88, 95, 176, 223–228, 250, 257, 264
   Zeh and, 30, 82–102, 104, 136, 143, 212–213, 248, 261, 270
   See also quantum physics
quantum physics
   Bohr and, 22–38, 47, 53–54, 256
   classical physics and, 68–70
   Einstein and, 17, 23–25, 42
   explanation of, 16–17
   foundations of, 94
   general relativity and, 29–31, 188
   laws of, 22
   nature of, 254–258
   quantum reality and, 85, 94, 254–257
   thermodynamics and, 188
   understanding, 208–214
quantum reality
   classical reality and, 37, 40, 80–81, 98–101, 250–257, 264
   complementarity and, 52–53
   hidden reality and, 9, 13, 23, 39, 136, 281–282, 282
   many worlds and, 66–67, 261
   multiverse and, 212
   on-screen reality and, 250, 282, 282–284
   perspective and, 13–16, *14*
   projector reality and, 13–17, *15*, 32, 39–40, 52–53, 123–124, 163, 250, 271
   quantum physics and, 85, 94, 254–257
   space-time and, 94, 102, 220–221
   understanding, 13–16, 19, 32, 39–40

quantum systems
  entanglement and, 6–7, 38–52, 65–66, 71–96, 100–103, 128–129, 136, 168, 179–180, 186, 201–220, 236–260, 270
  objects in, 6–7
  properties of, 6–7
Quantum Theory (book), 48
quantum waves, 21–24, 27–28, 31–35, 35, 46, 61, 90
quantum weirdness, 7, 33–34, 40–41, 84–85
quarks, 5, 100, 194, 202, 210–211

Raphael, 133
Rattansi, Piyo, 166–168
reductionism, 5, 204, 207, 283
Regiomontanus, 154
Reisler, Donald, 76
"Relative State Formulation of Quantum Mechanics," 75, 80–81
relativity theories. *See* general relativity; special relativity
religion
  "all is One" and, 49, 51–56
  as rebellion, 104–107
  science and, 55, 107–111, 124, 133, 144, 162–180
  struggle for "One" and, 103–145
  "unknown One" and, 284–288
Renoir, Auguste, 181
Republic (book), 32
"Reunited," 173
Richter, Peter, 160
Riddle of the Universe, The (book), 186
Ritter, Johann Wilhelm, 182–183
Roger II, King, 130
Romanticism, 4, 67, 168, 176–188, 248
Rosen, Nathan
  entanglement and, 41–47, 78, 91
  EPR paradox, 42–47, 66, 71, 78, 91–94, 214, 237–241, 261
Rosenfeld, Leon, 43, 75–77, 95
Rovelli, Carlo, 201, 214, 228–229
Rowen, Herbert H., 164
Royal Institution, 184
Royal Society, 149, 163–165, 170, 184, 288
Rudolf II, Emperor, 157, 159

Rumi, Jalal al-Din, 122
Ryu, Shinsei, 243

Sapiens: A Brief History of Humankind (book), 197
Saunders, Simon, 220, 227, 229
Savonarola, Girolamo, 139
Schaffer, Jonathan, 200–202
Schelling, Friedrich, 128, 174, 177–187, 284, 290
Scherrer, Paul, 21
Schiller, Friedrich, 173
Schiller, Jena, 177
Schlegel, August, 177–178
Schlegel, Caroline, 177–178
Schleiermacher, Friedrich, 164
Schlosshauer, Maximilian, 90, 257–258
Scholz, Hans-Joachim, 160
School of Athens (book), 133
Schopenhauer, Arthur, 51
Schrödinger, Anny, 18, 21, 198
Schrödinger, Erwin
  biology and, 198–199
  entanglement and, 47, 91
  equation by, 29, 36–37, 61, 69, 102, 224, 264
  illustration of, 9
  parallel universes and, 71
  philosophy and, 107
  projector reality and, 15–16
  quantum mechanics and, 18, 21–25, 28–37, 40, 43, 46–47, 53–56, 61, 86–87, 92, 163
  Schrödinger cats and, 34, 40–41, 65, 85, 89, 99, 239, 266–268
  waves and, 21–29, 36–37, 224–226
Schrödinger cats, 34, 40–41, 65, 85, 89, 99, 239, 266–268
Schrödinger equation, 29, 36–37, 61, 69, 102, 224, 264
Schrödinger's wave, 21–29, 36–37, 224–226
Schwarzschild, Karl, 232–233, 237, 242
Schwarzschild radius, 232–233, 237–238, 242
Sciama, Dennis, 78
Scientific American (magazine), 4, 160
Seiberg, Nathan, 247

self, understanding, 246, 249–271, 273–290
Seven Brief Lessons of Physics (book), 228
Seventh Letter (book), 113
Shakespeare, William, 166, 177
shamanism, 49, 110
Shelley, Mary, 182–183
Shelley, Percy, 183
Shimony, Abner, 94, 96
Short Treatise on God, Man and His Well-Being (book), 162
Sidereal Messenger (book), 157
Sidney, Philip, 166
Simons Foundation, 244
Singh, Ashmeet, 214
Socrates, 112–113, 128, 149
Sommerfeld, Arnold, 23
Sophia Charlotte, Queen, 169
Sophocles, 112
space
 defining, 8, 36–37
 entanglement and, 219–248
 quantum cosmology and, 219–231
 quantum reality and, 94, 102, 220–221
 time and, 7–8, 12–13, 36–37, 81, 91–94, 102, 126–128, 168, 195, 199–201, 210–214, 219–248, 252–254, 261–263, 269, 274–279, 285, 290
 warping, 223–225, 232
space-time geometry, 224, 231, 238, 242–243, 285
special relativity theory, 29, 44–45, 58–59, 220, 223
speed of light, 5, 44–46, 71, 252
Spenser, Edmund, 166
spin, 44, 94, 193, 202, 214–216, 270, 276–277
Spinoza, Baruch, 149, 162–163, 167–172, 178, 186
Standard Model, 192–194, 213
stars, 1–6, 11, 108, 143, 149–151, 172, 223, 226, 234, 264, 278
Stoica, Ovidiu "Christi," 255
"Strange (Hi)story of Particles and Waves, The," 86
Strathern, Paul, 138–139
Strawson, Galen, 275
string theory
 black hole and, 235–241
 explanation of, 7, 107, 195–196, 207–209, 213–214
 multiverse and, 238–239
 pioneers of, 31, 47, 207
 understanding, 207–215, 219–220, 228, 247
Sully, James, 236
Sunday Times Golden Globe Race, 266–267
superpositions
 "baby universes" and, 239–240
 classical reality and, 250–253
 decoherence and, 87–91, 258–260
 entanglement and, 72, 215, 242–243, 270
 explanation of, 3
 many worlds and, 257–260
 quantum weirdness and, 33–34, 40–41, 84–85
 space-time and, 242–243
supersymmetry (SUSY), 193–194, 211
Susskind, Leonard, 31, 47, 207, 213, 234, 238–240, 248, 252
Swingle, Brian, 243–244
symmetry, 57, 150, 188–189, 193–218, 231, 246, 255
Symposium (book), 128–129, 149

Takayanagi, Tadashi, 243
"Tao," 51–52, 54, 110, 114
Tao of Physics, The (book), 54
Tao Te Ching (book), 51, 110, 114
Tata, Xerxes, 284
Tegmark, Max
 bird perspective and, 97–98, 102, 213–214, 222
 consciousness and, 251, 255, 262–263, 270
 decoherence and, 97–98, 102, 128, 258
 information processing and, 278–279
 many worlds and, 238
 quantum factorization and, 251, 255, 262–263
 space-time and, 261–263
Teller, Edward, 10
"Temple of Nature," 185
Thales of Miletus, 107, 110, 114
Theodoric of Freiberg, 131
Theodoric the Great, 120

Theon of Alexandria, 118
"Theory of the Universal Wave Function, The," 75, 80–81
"There Are No Quantum Jumps, nor Are There Particles!," 86
thermodynamics
  black holes and, 233, 240–241, 263, 275
  entropy and, 205–207, 206, 229–233, 241–242
  quantum effects and, 89
  quantum physics and, 188
Thirty Years' War, 142, 159
Thoreau, Henry David, 181
Thorne, Kip, 10–11, 70, 233
thought experiments, 30, 34, 269–271
Timaeus (book), 112, 114, 154, 199, 277
time
  defining, 8, 36–37
  entanglement and, 219–248
  entropy and, 229–233
  lost time, 109, *219*
  quantum cosmology and, 219–231
  quantum reality and, 94, 102, 220–221
  space and, 7–8, 12–13, 36–37, 81, 91–94, 102, 126–128, 168, 195, 199–201, 210–214, 219–248, 252–254, 261–263, 269, 274–279, 285, 290
  timeless universe, 227–229
  warping, 223–225, 232
time travel, 13, 238
Toland, John, 169–170
Tononi, Giulio, 256, 261–263
Toscanelli, Paolo dal Pozzo, 135, 137–138
Trnka, Jaroslav, 244
True Intellectual System of the Universe, The (book), 167
Tucker, Alfred, 68
Turing, Alan, 276
Turing Machine, 276
Turner, William, 181
Tutankhamun, 50–51

"uglyverse," 194–197, 212
Ulam, Stan, 10
uncertainty principle, 25–28, *27*, 43–45, 50
Unity of Nature, The (book), 48

universe
  baby universes, 2, 195–197, 212, 239–240
  Big Bang and, 2, 13, 278
  branches of, 2, 72–73, 78, 78–80, 85, 99–100, 195, 239, 259–260
  clockwork universe and, 165–169, 175–176, 181
  description of, 1–6
  evolution of, 12
  history of, 12
  inflation of, 2, 195, 212–215, 238–239
  landscape of, 195, 201, 207–212, 238
  many worlds and, 2–3, 61–62, 66–67, 71–85, 89, 96–101, 142, 195, 212–213, 224, 238–239, 257–261
  multiverse and, 2–3, 71–77, 78, 78–85, 194–197, 212–213
  as "One," 3–8, 52, 96–97, *97*, 101–104, 113–114, 122–123, 274–275, 285–286
  one with, 1–8
  oneness of, 136–137
  parallel universes, 2–3, 71–72, 99–100, 238–239
  particles and, 4–5, 16, 21–24, 29, 33, 35–37, 282–289
  timeless universe, 227–229
  "uglyverse" and, 194–197, 212
  understanding, 1–12
  vastness of, 1–4
"Universe from a Single Particle, The," 246
University of Amsterdam, 240
University of Berlin, 17
University of British Columbia, 242
University of California, Berkeley, 83
University of California, Irvine, 260
University of California, San Diego, 83
University of California, Santa Barbara, 243
University of Cambridge, 166, 214
University of Chicago, 36, 267
University of Edinburgh, 169, 179
University of Göttingen, 18, 20, 57–58
University of Hawaii, 284
University of Heidelberg, 83, 135, 163
University of London, 51
University of Maryland, 240

Index

University of North Carolina, 74, 224
University of Padua, 135
University of Paris, 132–134
University of Pittsburgh, 264
University of Southern California, 79
University of Wisconsin, 256, 261
"unknown One," 273–290
Unruh, William, 88, 233
Unruh effect, 88, 233–234

Vaidman, Lev, 214
van Enk, Steven J., 215
van Gogh, Vincent, 181
Van Raamsdonk, Mark, 242–244
Vanderbilt University, 285
Vanini, Lucilio, 143, 169
Varela, Francisco, 268
Vedral, Vlatko, 215
Veiled Reality (book), 52, 163
Verlinde, Erik, 240–242
Vietnam War, 83
Vitruvian Man, The (drawing), 152
"von Neumann chain," 257
von Neumann, John
   "chain" and, 257
   entanglement and, 207
   EPR paradox and, 91
   quantum mechanics and, 34–35, 58–63, 66–71, 95, 256–257
   wave function and, 70–71
von Weizsäcker, Carl Friedrich, 48, 57, 62, 114, 201, 276

Wallace, David, 79, 100
Watson, James, 199
wave function
   collapse of, 33, 36, 44–46, 66–71, 97–98, 257–258
   equation for, 11, 224–228
   evolution of, 69–71, 98
   interpretation of, 33, 43–46, 52, 66–81, 97–98, 224–226, 231, 246–258, 281–282
   quantum mechanical wave, 86–90, 97, 136, 224–225, 257, 281
   universal wave function, 69–81, 87, 224–226, 246–252
"Wave Function: It or Bit," 281

waves
   particles versus, 16, 21–24, 29, 32, 55
   properties of, 16, 18, 21–24, 33
   quantum waves, 21–28, 31–35, *35*, 46, 61, 86–90, 97, 136, 224–225, 257, 281
   Schrödinger's wave, 21–29, 36–37, 224–226
Weiler, Tom, 285
Weyl, Hermann, 21, 35, 62, 222
Wheeler, Joe, 10
Wheeler, John
   illustration of, *12*
   "It from Bit" and, 276, 281
   many worlds and, 72–76, 80–81
   projector reality and, 13–17
   quantum gravity and, 69–74, 224, 233
   quantum mechanics and, 10–13, 28, 37–38, 52, 63, 66, 77, 82–83, 88, 95, 176, 223–228, 250, 257, 264
   Wheeler-DeWitt equation, 11, 224–228
   "Wheeler's U," 10–13, *12*, 37–38, 176, 250, 264
Wheeler-DeWitt equation, 11, 224–228
"Wheeler's U," 10–13, *12*, 37–38, 176, 250, 264
Whitaker, Andrew, 92
White, Martin, 160
Whitehead, Alfred North, 112
Whiteley, Giles, 182
Whitman, Walt, 181
Wien, Wilhelm, 23–24
Wigner, Eugene, 63, 95, 257–258
Wilczek, Frank, 188, 221–222
William III, Prince, 164
Witt, Johan de, 164
Witten, Edward, 213–214
Wittmann, Marc, 268–269
Wolchover, Natalie, 197, 216, 245
Wordsworth, William, 4, 165, 180–181, 184
Works and Days (book), 108
World Pantheist Movement, 286
World Until Yesterday, The (book), 49
World War I, 18, 232
World War II, 10, 67, 82, 276
wormholes, 11, *219*, 237–240, 242

Xenia (book), 174
Xenophanes, 110–111

York, Michael, 49

Zarlino, Gioseffo, 156
Zeh, H. Dieter
    decoherence and, 84, 87–97, 99–101,
        128, 136, 227–231, 250, 258–260
    entanglement and, 52, 91, 136, 213
    illustration of, 65

projector reality and, 16
quantum mechanics and, 30, 82–102,
    104, 136, 143, 212–213, 248, 261,
    270
time and, 230–231
wave function and, 281
Zeilinger, Anton, 90, 93
Zini, Modj Shokrian, 246
Zurek, Wojciech, 77, 81, 90, 96,
    251–254
Zuse, Konrad, 276

Heinrich Päs is a professor of theoretical physics at TU Dortmund University in Germany. He earned his PhD from the University of Heidelberg for research at the Max Planck Institute, has held positions at Vanderbilt University, the University of Hawai'i, and the University of Alabama, and has conducted research visits at CERN, Fermilab, the Gran Sasso Laboratory, and more. Päs has written a *Scientific American* cover feature, and his research has been featured three times on the cover of *New Scientist*. He lives in Bremen, Germany.

Photograph by Emilia von Dombrowski and Bogumil Longowski